The Experiment of Life

Science and Religion

Edited by F. Kenneth Hare

This volume originated in a conference held at Trinity College, University of Toronto in 1981 to mark the 100th anniversary of the birth of William Temple, who was Archbishop of Canterbury and an influential Anglican philosopher and theologian. Seven contributors – John Macquarrie, D.R.G. Owen, A.R. Peacocke, Kenneth E. Boulding, P.W. Kent, Alan M. Suggate, and Robert W. Kates – review the present condition of the frontier (or no man's land) between science and religion. Temple's insights into this area are debated, but the main thrust of the authors is to look at the modern world as he might have done.

Recent discoveries in molecular biology and the neurosciences have raised major questions about human conduct and extended the mechanistic viewpoint into this realm. The interdisciplinary science of sociobiology offers a new and controversial view of behaviour – and hence of morality. Examining the theological and the scientific traditions of thought, these authors ask: Are they reconcilable? On the whole their conclusions are optimistic, but a greater effort is called for to lower the artificial barriers put in place by both the organization of these traditions and the compartmentalizing structure of the universities.

Another series of issues – stemming from mankind's misuse of the environment – is debated in depth. Questions of social ethics are treated both as Temple saw them and as they now appear in the light of a half-century of world conflict. Social and environmental issues, however, force the debate of science and religion out into the public gaze. Though the sciences are widely used in the analyses of these problems, ethical dilemmas repeatedly emerge that our society cannot avoid, yet has difficulty resolving.

This book provides a series of commentaries on such matters of concern. All are based on long intellectual involvement and personal commitment. Each article is self-sufficient, yet relates to the others around Temple's unifying theme: that we inhabit a sacramental universe, and that life is the grandest experiment.

F. KENNETH HARE is Provost of Trinity College, University of Toronto

CONTRIBUTORS

Kenneth E. Boulding
University of Colorado, Boulder

F. Kenneth Hare (editor)
Trinity College, University of Toronto

Robert W. Kates
Clark University, Worcester, Massachusetts

P.W. Kent
Van Mildert College, Durham, England

John Macquarrie
Christ Church, Oxford, England

D.R.G. Owen
Trinity College, University of Toronto

A.R. Peacocke
Clare College, Cambridge, England

Alan M. Suggate
University of Durham, Durham, England

The Experiment of Life

Science and Religion

Edited by F. Kenneth Hare

Published in association with the
University of Trinity College, Toronto
by
University of Toronto Press
Toronto Buffalo London

Toronto Buffalo London
Printed in Canada

ISBN 0-8020-2486-6 (cloth)
ISBN 0-8020-6506-6 (paper)

Canadian Cataloguing in Publication Data

Main entry under title:
The Experiment of life : science and religion

Proceedings of a conference held at Trinity College,
Toronto, Aug. 1981.
ISBN 0-8020-2486-6 (bound). – ISBN 0-8020-6506-6 (pbk.)

1. Religion and science – Congresses. 2. Temple,
William, 1881–1944 – Congresses. I. Hare, F. Kenneth
(Frederick Kenneth), 1919– II. Trinity College
(Toronto, Ont.).

BL241.E96 261.55 C83-094087-1

Contents

Introduction: The Experiment of Life – and Beyond

F. KENNETH HARE

This volume contains most of the papers read before a conference held at Trinity College, Toronto, in August 1981 – one hundred years after the birth of William Temple, Anglican divine, ecumenist extraordinary, and scholar of rare qualities.

The event was planned as an academic conference on science and religion, with the title 'The Experiment of Life.' These were the last four words of Temple's Gifford Lectures to the University of Glasgow, given in 1932 and 1933–4, almost fifty years ago.[1] The centenary of his birth seemed a good time for a hard look at the relations between science and religion in the modern world, and especially in the university. As it happened the discussion was even broader, spilling over into social ethics. It touched on Temple's entire intellectual achievement, and went beyond it. In fact the debate was not so much about Temple as about the modern situation as it might have struck him.

Perhaps this outcome was just as well. It is true that Temple was an academic philosopher before he became a priest. Though he was one of the great reformers and scholars of his day, a main thrust of his life was nevertheless still religious and ecclesiastical. If his centenary was to be celebrated effectively, today's churchmen had to have their place on the agenda. As it turned out they seized the opportunity, both on the platform and in the audience. There were many more parsons than professors on the floor of the house. And the debate, when it became critical, was directed towards the failings of the churches, not the weaknesses of the universities.

It was from churchmen that the idea for the conference came in the first place. Trinity College, though a secular member of the federal University of Toronto, is a foundation of the Anglican Province of Ontario. Bishop George Snell, fomerly bishop of Toronto, had long admired Temple's life and works. He suggested to the metropolitan of Ontario, Archbishop Lewis Garnsworthy, that the Diocese of Toronto should celebrate Temple's centenary birthday – 15 October 1981. A festival in Temple's honour was in fact organized by the diocese. As its contribution Trinity College agreed to organize a major conference that would consider the world of today as Temple might have seen it.

Several of us had known or at least met Temple, and most of us shared Bishop Snell's admiration. For myself, I remembered Temple as an Olympian figure who had suddenly and mysteriously descended upon meetings of a society that I joined while an undergraduate at King's College London, of which Temple was chairman of council. He was also archbishop of York and primate of England at the time. I was mildly staggered that a prelate of such eminence should want to help some puny undergraduates who were worried about the welfare of animals. But in fact he did. Others among my Trinity colleagues had met him on more familiar terrain. None could have been more impressed than I was. Temple was a towering but remote figure to those of us who were undergraduates in the 1930s – a figure of the high establishment who nevertheless appeared as a prophet of social reform, and who cared about time-expired horses. There was no one like him in my fledgling days.

And so in preparing the conference we asked ourselves: What part of Temple's work should we look at? The complete corpus was far too broad. We suspected that others would seize on his treatment of the Fourth Gospel, and on his views on pastoral care and church life. His role as an ecumenical figure, especially in relation to the formation of the World Council of Churches, seemed to broad and too ecclesiastic for us to tackle – though the present moderator of the

World Council, Archbishop Scott, primate of Canada, is in many ways Temple's successor. We decided to start from *Nature, Man and God* and let our speakers define the content. The science–religion debate would be our focus, but not to the extent of restricting freedom of argument.

Debate may be too strong a word. In the universities there is little real talk between natural scientists and those who study religion, whether theologians, philosophers, historians, or even anthropologists. Those who *do* talk write many books and articles. There is an extensive literature about science and religion, and the bibliographies have many recent entries. There are many specialists on this subject, some of whom came to the conference. For those of us who have spent our lives in the universities, however, the frontier between these two traditions is still a no man's land, crossed only here and there by the hardy souls who persist in thinking that the two should at all costs be held together. For the most part this frontier is an empty intellectual strip across which not even shots are fired, a truly demilitarized zone. Faculties of science and divinity can share such a frontier with the utmost politeness. There is no longer much fear of ill-bred arguments to disturb their peaceful coexistence, still less to challenge their intellectual credentials.

This isn't a bad image of the modern university. Reductionism has had its say, and nineteenth-century German ideas about organized disciplines have compartmentalized the institution. It would, in fact, be more apt to talk about *com*partments than *de*partments. Few outsiders realize the extent to which this compartmentalization has robbed the university of any capacity for collective judgments on things that matter. For my sins I have been head of one university and two colleges and dean of a large faculty at another. In all these capacities I was often asked the opinion of the body I headed about some substantive matter. Always I had to say that it had no opinion. Universities (and even colleges) nowadays confine their collective judgments to matters of logistics, administration, money, and public relations. Meetings of academic bodies confine their agendas to trivia. The tradition has become one of departmental or even personal privacy: you don't really challenge your neighbour. Up to a point this attitude is good. It makes life pleasant. But it also means that the greater intellectual issues will be talked about mainly off the campus. The university is a house of many mansions without a good intercom system, and with discreetly closed shutters.

And surely this is one of those greater issues. Religion tries to go to the ground of our being. Both science and religion deal with the nature of things, as does (or did) philosophy. It is the largely silent conviction of most scientists that the only way to understand the ground of our being is to pursue the methods that we now call scientific – a mixture of empirical observation, reasoning, and detached judgment that is much the same when applied to textual criticism of the Bible as to the analysis of the DNA molecule. It is the largely unvoiced conviction of these sceptics that assertions with any other basis are in fact meaningless. (This stance expresses the positivism for which they are both rightly or wrongly attacked.) How, then, can the theological lamb sit down with the scientific lion, who privately thinks that the theologian is wasting his time? And how can the scientist ignore the confident assertions of the theologian about the existence of God, whom Laplace found an unneeded hypothesis? The answer is simply that the organization of the modern university makes such challenges embarrassing, untimely, and (for the challenger) pretty risky. In the modern university you need allies, not enemies. And sometimes those allies are useful in the schools, too, from which the universities get their students. In the creationist assault on science teaching in North American schools the strongest defenders of the validity and integrity of science have been churchmen and theologians.

Obviously science was too broad to be tackled comprehensively at the conference. In any case the Vatican had dealt with the physical sciences. In 1979 Pope John Paul II had convened an ecumenical seminar of the Pontifical Academy of Sciences, and the participants had chosen to re-examine the work of Einstein. Modern physics has been kind to the theologians. It has shed the dogmatic certainties of nineteenth-century mechanics and thermodynamics. The published papers, by Carlos Chagas, P.A.M. Dirac, and Victor F. Weisskopf, displayed a gratifying consciousness that the physical universe is not and cannot be dogmatically known.[2] It is a staggering place, and there might even be room for God (at least none of the speakers ruled him out). Impressed by the new generosity of the scientists, Pope John Paul took the first step towards the rehabilitation of Galileo – a gesture that may well have surprised the delegates.[3] So the long domination of the science–religion debate by those who sought to reconcile faith with physics was alive and well in the halls of the Vatican.

We lacked the Vatican's resources, and had no urge to carry on the same argument. It seemed to us that there was a more vital topic – the experiment of life, as Temple put it. The suggestion came originally, I think, from Bruce Alton of Trinity College, Toronto. It was and is my own view that the phenomenon of life, and the discoveries that science has now made about its physics and chemistry, are central to the science–religion controversy. They should be the main topic of the debate in 1981. We decided to limit our subject that way. The coincidence that Teilhard de Chardin's centenary also fell in 1981 did not influence our choice. But at least it was a notable coincidence.

We had read, of course, the work of Francis Crick, in which he announced confidently that modern biology aimed to explain '*all* biology in terms of physics and chemistry.'[4] Equally we had been struck by the vigorous rebuttal by Arthur Peacocke, biochemist and priest, who deployed the term 'nothing-buttery' for the view that life is nothing but physics and chemistry.[5] We asked Peacocke to be a keynote speaker at the conference. His paper, published in this volume, goes well beyond his earlier work, and reviews compositionist aspects of the new biology, such as evolution and ecology, as well as the thoroughly reductionist topics of molecular biology and sociobiology.

There have been other developments in the science of life that touch on the science–religion debate. The neurosciences and modern genetics needed spokesmen, and we asked William Tatton and Louis Siminovitch to be those men. 'I see no room for free will,' said one of them prior to the conference, and some of us wondered then and afterwards whether there was room for morality in a world that lacked free will. Environment, too, was a question we could not avoid. Robert Kates was our spokesman on the place of man in nature, ably supported by Ian Barbour. His paper, published here, raised many issues that are inescapably ethical, and which were obviously novel to many churchmen.

We failed, however, to find any real 'nothing-butters,' partly because they are not all that numerous, but mainly because hardliners avoid such debate. One finds them in most fields of science. They are men and women who go beyond the ordinary working methods of science to assert, often in their own minds only, that there really is nothing in the universe but mindless mechanics. They present the hard face of modern science. They are unimpressed by the argument that an external Creator might well be expected to maintain a highly consistent system of mechanics to achieve Creation. We failed to produce spokesmen for any part of hard-faced science; all the scientists who came were studiously polite and open-minded, though scarcely vocal. They listened, but mostly didn't comment.[6]

It was left to an economist, Kenneth Boulding, to speak for science as a whole. Boulding sees science in terms of its history, culture, and fallibilities. His presidential address to the American Association for the Advancement of Science was an electrifying event that packed the halls of the San Francisco Hilton, and then rocked the roof with approving laughter.[7] Boulding's unique achievement in the modern world is that he makes us think most acutely when we laugh our hearts out. The inner logic of science, he argued, called for a 'decent agnosticism.' The biology textbooks, he said, needed to be 'a little more modest, saying things may have happened this way, but we are not quite sure that they did.' He also felt that 'we should not be over-confident or chatty about the large design of things.' And that was his listeners' view too, as they chuckled away.

Last among the scientific speakers was Paul Kent, like Peacocke a biochemist. He talked about creativity in science, pre-eminently an activity that withers unless it has new ideas to feed on. He spoke of the growing fear of science, and of the need for scientists to extend themselves to justify their work to the public at large. Temple, Kent reminded us, was not closely identified with the science–religion debate of his day. Science today might be an abstraction from the real concerns of mankind, or even a threat; if society's knowledge outgrew its wisdom, Kent argued, danger might exist for us all.

To put this discussion in its proper setting John Macquarrie began with an encapsulation of Temple's life and work as an intellectual. He treated Temple as philosopher, theologian, and churchman, making it clear that one could not confine Temple's relevance for the late twentieth century to his philosophy and critique of science. It was not *Nature, Man and God* that was now widely read, but *Christianity and Social Order*,[8] published during the disorder of the Second World War. To Macquarrie Temple's social teaching might seem, in 1981, to be 'mild and reformist' – which, Macquarrie said, the Christian was bound to be. Later, in a public address to be published elsewhere, he developed a view of Christian social ethics that went far beyond what I can detect in Temple's work.

Macquarrie's words led into a paper from Alan Suggate, who developed two themes – a critique of Temple's own ethical approach, and a personal position on Christian social ethics and its future. Like Macquarrie, Suggate went far beyond Temple, whose last twenty years were spent 'juggling rather untidily with several different approaches.' It was a sign of Temple's greatness, according to Suggate, 'that he is most interesting at the end of his life.'

It was striking, I thought, how the temperature of the debate went up as these papers were presented. The listeners had enjoyed the scientists, and had largely accepted their propositions. There were few challenges. But there was also no passion, no anger. Only when we progressed to the social and ethical issues did those emotions show themselves. As I said, we were far more parsonical than professional. For parsons it was clear that these issues of ethics lay closer to their own solar plexi. We were wrong not to have foreseen this reasonable bias.

Yet the most striking phrase in all Temple's work is *the sacramental universe*.[9] It was this famous Gifford lecture that in many ways foreshadowed Peacocke's position, reinforced by the views of another giant whose teaching is less visible today than it should be – Michael Polanyi.[10] It was to this sacramental view of reality that the honorary chairman of the conference, Derwyn Owen, addressed himself. Owen reminded us (as did Peacocke) of the solid materialism of Temple's teaching, so much more palatable to the scientist of today than the Cartesian position. Without Owen's address I might have been persuaded that mankind is the proper and sufficient study of the humanistic scholar. But the Word was made flesh, and dwelt among us. And it dwelt, as we dwell, in a physical universe. Christians are often accused, rightly as I think, of undervaluing nature. It is no service to ourselves to ignore the natural world around us.

What, then, did we achieve? Were the four days well spent?

We concluded, I believe, that science was indeed too important – too *damned* important – to be left to the scientist. We couldn't teach him how to do his work, or to refine his methods. But we could (and did) demand that he relate his work to the larger body of human achievement, and be patiently accessible to all others who want to understand reality. We clearly felt that he could not escape responsibility for the social functions of science. To do him justice, he usually didn't want to.

Reductionist science, we acknowledged, was a powerful and necessary weapon in the intellectual armoury. It had led in our own time to a dramatic revelation of the mechanics of biological development, inheritance, and evolution. It offered new insights, as yet controversial, into theories of human behaviour. Conceivably the ethologist had something to say about ethics. But even within science reductionism was insufficient. Evolution and ecology required what Peacocke called compositionist approaches. There were many aspects of science that required broad, interdisciplinary learning, strongly akin in some ways to good humanist teaching.

Such interdisciplinarity was hard to achieve, because it meant more than putting together teams of specialists. Harlan Cleveland once said that interdisciplinary courses in a university were always given by such teams, so that only the students had to be interdisciplinary.[11] Yet somehow the scientist has to create holistic frameworks, for in nature it is often the case that the whole is greater than the sum of its parts.

We were far less agreed on the role of theology, and in my final summing-up I was obliged to put my own views forward. They were received with good-natured neutrality. I asserted that theology, too, was far too important to be left to the theologians. I suspect that this proposition was accepted out of a sense of fair play: if we were going to be hard on the scientist, we had to be just as hard on the theologian.

I submitted, without being challenged, that the literate public demanded a consistent and defensible picture of God, stripped of sentimentality and pieties, and satisfying to the intellect as well as the heart. If not, why did the quality press devote so many column centimetres to the battle between Hans Küng and the Vatican? Why did the suffragan bishop of Woolwich become a best-selling author with *Honest to God*?[12] And why did a heated rebuttal called *For Christ's Sake*[13] sell so many fewer copies? Clearly because the image of God still matters to a public that has largely written off the churches. The conference was reminded of Collingwood's idealist history of the 1930s and 1940s, in which we were invited to rethink the thought of the past – and of G.J. Renier's question: 'How are we to prepare for this exquisite symbiosis except by going into a trance?'[14] What the public needed was an intellectually consistent picture of God and his world, without the use of such narcosis.

And in assembling that picture it was clear that the modern theologian could not avoid a contemplation of life, of the world, and of the universe as revealed by observation. The universe was a staggering, almost inconceivable entity – perhaps not even an entity. Early training as a physical scientist led one to expect that in the end – one used to talk grandly about 'the ultimate synthesis' – all would be resolved into a few general principles. Fundamental simplicity would be revealed even behind the chaos that the second law of thermodynamics predicted. But now things seem less tidy. The physical universe is anything but simple, and the physicist (while still seeking simplicity as earnestly as did Occam) knows that comprehension of nature calls for gigantic exertions.

A theology that will face up to this gigantic challenge belongs in a university, and not just in isolated seminaries. At Trinity we are lucky enough to have faculties of divinity and arts alongside one another. They are not separated by high walls and *cordons sanitaires* as they are in many places. This juxtaposition is an inheritance from the founder, Bishop John Strachan. I hope that in due course we shall profit more from it than we do today. Equally I hope that in other places and at other times new juxtapositions will arise. If God exists, he will be impossible to encompass, and very difficult to be knowledgeable about. Keeping theologians close to other humanists and to the scientists is a good way to help things along.

Repeatedly, as we planned this conference, I was reminded that Temple is no longer a name to be conjured with. Churchmen, reformers, and philosophers have gone to other things. He looks dated today. But the Trinity conference, small and compact as it was, showed that this view is itself dated. By celebrating the hundredth anniversary of his birth, we found that some of his concerns are still as vital as ever. 'We are so impressed by the greatness and multiplicity of the world we know,' he wrote, 'that we seldom reflect upon the amazing fact of our knowing it.'[15] I'm not sure that he shouldn't have written 'knowing *of* it.'

In any case we disbanded in good heart. I hope that the published papers will intrigue others, and push the world marginally closer to understanding. Of what, the reader must judge.

NOTES

1 William Temple, *Nature, Man and God*, being the Gifford Lectures delivered in the University of Glasgow in the academical years 1932–1933 and 1933–1934 (London: Macmillan 1934)

2 The addresses, as well as that of Pope John Paul, were published in English in *Science* 207 (1980): 1159–67.

3 'I hope that theologians, scientists and historians, imbued with a spirit of sincere collaboration,' said the Pope, 'will more deeply examine Galileo's case, and by recognizing the wrongs, from whatever side they may have come, will dispel the mistrust that this affair still raises in many minds, against a fruitful harmony between science and faith, between the Church and the world.' (*Science* 207 [1980]: 1166)
4 F.H.C. Crick, *Of Molecules and Men* (Seattle and London: University of Washington Press 1966), 10
5 The phrase occurs on p. 32 of this volume.
6 The idea that a consistent, invariably mechanical universe is proof, and not disproof, of an external creator no doubt has many origins. I first heard it from David Brunt, the physical secretary of the Royal Society of London.
7 K.E. Boulding, 'Science, Our Common Heritage,' *Science* 207 (1980): 831–6
8 W. Temple, *Christianity and Social Order* (Harmondsworth, UK: Penguin 1942). Reprinted, with an introduction by Ronald Preston (London: SPCK and Shepheard-Walwyn)
9 Lecture 19, the Gifford Lectures (1933–4), 473–95 (see note 1)
10 For example, *Personal Knowledge* (London: Routledge and Kegan Paul 1958)
11 Harlan H. Cleveland, in *Science and Future Choice*, ed. P.W. Hemily and M.N. Ozdas, 2 vols. (Oxford: Clarendon Press 1979), 2: 343–4
12 John Robinson, *Honest to God* (London: SCM Press 1963)
13 O. Fielding Clarke, *For Christ's Sake* (Wallington, UK: Religious Education Press 1963)
14 G.J. Renier, *History – Its Purpose and Method* (Boston: Beacon Press 1950), 48
15 Temple, *Nature, Man and God*, 129

The following are major works of William Temple that provided a basis for discussion of the principal issues of this conference:

Christianity and Social Order. Harmondsworth, UK: Penguin 1942
Christianity in Thought and Practice. New York: Morehouse 1936
Christus Veritas. London: Macmillan 1924
The Faith and Modern Thought. London: Macmillan 1910
Nature, Man and God. London: Macmillan 1934
Readings in St. John's Gospel. London: Macmillan 1939

The Experiment of Life

William Temple: Philosopher, Theologian, Churchman

JOHN MACQUARRIE

Born 1881; died 1944. Just to quote the dates of Temple's life is to say quite a lot about him. In 1881, Queen Victoria was on the throne and only six years away from her Golden Jubilee, Mr Gladstone was prime minister for the second time, and the British Empire was in its full glory. In 1944, the Second World War had been raging for five years, the superpowers were the United States and the Soviet Union, and the England into which Temple had been born was in the process of radical transformation. Temple's life spanned the period of transition from what seemed the unshakeable structures of late Victorian times to the uncertainties and anxieties of the modern world.

But he himself was no mere spectator of the vast changes that were going on. He was an active participant. He was influential in the world of thought, particularly in philosophy and theology; he played a major part in ecclesiastical affairs, not only in the Church of England but in the ecumenical movement; but perhaps he will be chiefly remembered for his work for society at large – in education, in politics, and in efforts to reshape the economic order. One gets the impression that very few churchmen ever attain to an influence over politicians. Temple was one of those few. Edward Heath has written about him: 'The impact of William Temple on my generation was immense ... The reason was not far to seek. William Temple was foremost among the leaders of the nation, temporal or spiritual, in posing challenging, radical questions about the nature of our society.

Most important of all, he propounded with lucidity and vigour his understanding of the Christian ethic in its application to the contemporary problems which engrossed us all.'[1]

But let us go back briefly to 1881. Temple was born into the Establishment – one might say, with a silver ecclesiastical spoon in his mouth. He saw the light of day in the episcopal palace at Exeter, where his father, Frederick Temple, had been appointed bishop, not without some controversy, a few years earlier. His father, of course, went on to be bishop of London, when William was four years old, and then to be archbishop of Canterbury. William received the education considered proper for a member of the English upper class, that is to say, public school and ancient university, in his case, the impeccable credentials of Rugby and Balliol.

From then on, his career reads like the ecclesiastical success story *par excellence* – it would have been the envy of even the most ambitious dignitary of Barchester Towers. Fellow and tutor in philosophy at The Queen's College, Oxford; headmaster of Repton; rector of St James, Piccadilly; canon of Westminster; bishop of Manchester; archbishop of York; and finally, archbishop of Canterbury. It might seem a pity that he had to take in such an ugly city as Manchester in his triumphal progress. Yet it may have been his eight years of ministry in Manchester that lay nearest to his heart. If we look at some of the things he was doing 'on the side,' as it were, rather than at his list of official appointments, we get a very different picture. Already while a layman he was active in the Workers' Educational Association. For seven years he was a member of the Labour party, though he left it in 1925 and what he would have made of it today is hard to know. He organized conferences on social questions, especially a notable conference on Christian Politics, Economics and Citizenship, held at Birmingham in 1924. According to Edward Norman, this conference 'marked the rise of William Temple to ascendancy over the social teaching of the Church of England.'[2] Meanwhile he was also busy with ecumenical conferences of 'Faith and Order' and 'Life and Work.' Within the Church of England, he was chairman for several years of a commission on doctrine. Its report, published in

1938, was one of the best things that have been done on this subject, and is still being quoted. Temple's relatively early death prevented him from ever presiding at a Lambeth conference, but as archbishop of York he was one of the most active figures in the Lambeth Conference of 1930. After an unimpressive beginning, he made major contributions, especially in the area of ecumenical relations.[3]

I have been dwelling on Temple's achievements as social reformer and churchman, but one would be giving only half the picture by representing him as an activist. He would have despised those activists who repudiate theology and spirituality. He had, after all, begun his career as a philosopher. He wrote: 'Thought and action, if separated from one another, are abstractions and falsifications; the real fact is active rational life, and this is best expressed in rational activity, where thought and conduct are inextricably united.'[4] His devotion to learning is perhaps best attested by the story that he spent the night before his wedding finishing his philosophical book, *Mens Creatrix*. The most striking thing about Temple's thought is its extraordinary range. He had a remarkable mastery of philosophy, theology, science, and both English and classical literature. All of these elements flow together into his own synthesis – and synthesis is the word, for Temple sought to bring things together and to see them in their unity and wholeness. This quest for unity was no doubt part of the legacy he had received from his most respected teacher, Edward Caird, and even if he was later attracted to Whitehead, he continued to do philosophy on the grand scale.

Temple's philosophy is the obvious place to begin in any account of his thought, though his philosophy incorporates his science and flows into his theology, so that it would be impossible to make sharp divisions among them. Temple's idea of philosophy was virtually the opposite of that which still prevails today throughout most of the English-speaking world. That contemporary philosophy is analytical and critical in its style. Temple, by contrast, believed that it is the business of philosophy to show the unity and coherence of experience. He tells us that the standard of the mind in its search for truth is totality – 'the embrace of all relevant reality in a comprehensive unity,' to quote his own words.[5] Of course, the presupposition of this belief is that the world is a rational whole and that the principle of unity is graspable. It is perhaps a pity that although analytical philosophy was already in the ascendant in the years before Temple's death, he took virtually no interest in it. It is perhaps even more a pity that he did not interest himself in Kierkegaard, who was being rediscovered about this same time. I believe myself that philosophy has always two tasks, analysis and synthesis, but I should think that most contemporary philosophers coming to Temple would judge that there is too little critical analysis in his work, and that his claim to be building up a philosophy out of the facts discoverable in the various areas of experience and investigation is strongly suspect. One critic, Owen Thomas, declares: 'He was a man of profound Christian faith, and what he actually has done is to arrange the facts so that they give support to his faith.'[6] Perhaps Thomas should not have used the word 'arrange,' which conveys a hint of artificiality and manipulation. But I think his general point is justified, namely, that Temple's philosophy is to be seen as a form of apologetics. There is a tacit assumption of the truth of Christian theism, and then an exposition and interpretation of the facts of experience so as to show that they are highly compatible with the initial unspoken assumption. Of course, apologetics is a perfectly legitimate exercise, and Temple's work is in many ways skilfully done, but we should be clear what he is doing.

How is the philosophy underlying Temple's synthesis to be characterized? It is usually said that he moved from idealism to realism – in particular, from the influence of such thinkers as Caird, Bosanquet, and Royce to that of Whitehead. Broadly speaking, this statement is true, but the terms 'idealism' and 'realism' are notoriously vague and ambiguous, and Temple's synthesis retained elements of each.

There is no doubt that he was an epistemological realist; that is to say, he believed that we know things and facts in the real world, and not just ideas in our own minds. This is the point of his brilliant and justly famous criticism of Descartes: 'If I were asked what was the most disastrous moment in the history of Europe, I should be strongly tempted to answer that it was that period of leisure when

René Descartes, having no claims to meet, remained for a whole day "shut up in a stove."'[7] That day began the 'turn to the subject' in western philosophy. Only in the present century are we escaping from it, and Temple is in good company among modern philosophers in his rejection of Cartesianism. But we should notice that he also considered that Cartesianism constituted 'a necessary movement of thought.'[8] Temple retained from his idealist origins a belief in dialectic, especially in the history of ideas. The Middle Ages had been predominantly realist, so realism is the thesis. Cartesian and post-Cartesian philosophy was the antithesis that corrected the defects in medieval realism. But Temple believed that in dialectics the weight of truth lies with the original thesis, that the antithesis is secondary and dependent in character, and that in the final synthesis we shall be closer to the original thesis, though suitably modified to take account of whatever was valid in the antithesis. Thus he believed that we are now moving into a phase that will resume the epistemological realism of the Middle Ages but will also incorporate the 'turn to the subject.'

As well as epistemological realism, there is also metaphysical realism, with metaphysical idealism as its correlate, and here it is harder to determine whether Temple really broke out of idealism. It may well have been his respect for the natural sciences that encouraged him towards epistemological realism in the first place. With that basis he then had to accept the evolutionary scheme presented in the modern scientific picture of the world. This view implied that there was a time when mind had not appeared in the world, and matter existed independently of mind. Temple accepts this fact as a refutation of the Bishop Berkeley type of idealism, according to which what we call 'material things' are in fact percepts of mind – *esse est percipi*; their very existence consists in their being perceived. But then, as his argument develops, Temple tends to reverse his stance. Mind is said to be secondary and derivative in the process; it first manifests itself as the behaviour by which organisms cope with their environment, but then it is set free from immediate biological needs, it begins to apprehend and even to comprehend the process itself, and it eventually transcends the process and becomes the clue to the nature of the whole. To quote his words: 'It achieves a certain superiority to, and independence of, the process – not indeed such as to endow it with a life wholly detached from the process, but such that the process falls within its grasp, not it within that of the process.'[9] Are we not back finally to a kind of metaphysical idealism in which mind is the creative source of everything? Temple is, of course, correct in his contention that a process that brings forth minds capable of understanding and transcending it is not ultimately explicable in reductionist terms. But if one then posits that mind – indeed, a divine, creative, transcendent mind – has been there from the beginning, is it not misleading to claim to be a metaphysical realist and to hold that the ultimate reality is the emergent process that has been going on for billions of years before mind appeared? C.D. Broad, a near contemporary of Temple, represents a genuinely realist position in his book *The Mind and Its Place in Nature*, a kind of emergent materialism in which the mind does not have the privileged position that Temple gives it. Temple never really moves from the metaphysical idealism expressed in the title of his early book, *Mens Creatrix*.

The difficulties of Temple's position become clearer when we consider his ambiguous relation to Whitehead. Whitehead begins from the processes of nature as interpreted by modern science, but he is able to end with a metaphysic that finds room for value, spirit, and even God. The attraction of this view for Temple is obvious. Whitehead overcomes that bifurcation or dualism so disastrously introduced into Western thought by Descartes. But Whitehead's metaphysic is finally rejected as inadequate. Its highest category is that of organism and, according to Temple, we must go beyond that to the category of personality. But this judgment means also that Whitehead's God likewise is deemed to be inadequate, in the sense that he is immanent in and correlative with the process of the world. 'Personality,' declares Temple, 'is always transcendent in relation to process.'[10] Temple struggles long and hard to work out a satisfactory conception of the relation of God and the world, one that would avoid that correlation which, to quote his own words, 'we found to be the capital error of Whitehead's philosophy.'[11]

In more than half a dozen passages scattered through his Gifford Lectures we find Temple wrestling with the God–world problem. In most of these passages he is upholding the transcendence and independence of God, as against what he considers to be the 'capital error' of Whitehead, yet in each case he modifies what he has to say to allow some scope for divine immanence. In other passages, he affirms the immanence of God, the necessity of creation, and the value of history, yet draws back from a fully 'process' type of theology. He criticizes some analogies as making God too external to the world, others for correlating him with the world, but nowhere gives a clear statement of the middle way that he is seeking. So we find him saying that the world is dependent on God as he is not dependent on it. Yet he is not external to it, like a carpenter who has made a box, and not even as a poet who has composed a poem. That second analogy, he says, 'is an improvement on the analogy of the carpenter and his box, but it leaves God too external to the world.'[12] In another place he gives a far more extreme statement. He writes: 'Only by a doctrine of thoroughgoing transcendence is the world explicable or the religious impulse satiable. In the sense in which God is necessary to the world, the world simply is not necessary to God. Apart from him, it has no being; apart from it, he is himself in plenitude of being. The world − God = 0; God − the world = God.'[13] But he immediately modifies this view because it would deprive history of meaning, and history is important to religion, especially Christianity. Again we find him saying: 'Upon God, the world depends for its existence; in no sense does God depend on the world for his existence.'[14] But this statement, which might seem to encourage a kind of acosmism, is qualified by the assertion that the world's aspirations and heroisms have value for God, and especially the historical events associated with Jesus Christ.

But now we look at the other side of the balance sheet. He says that 'if all else but God were abolished, God would still be himself, whole and entire,' but he adds, 'capable of creating another world to take the place of the world that had gone out of existence.'[15] There is a hint here that some world is necessary to God. That word 'necessary' is introduced a few pages further on: 'God is active in the world, and its process is his activity. Yet he is more than this; he is creator and therefore transcendent. Because he is, and is creative, he must create; therefore the universe is necessary to him in the sense that he can only be himself by creating it.'[16] I do not see how this last sentence expresses a view any different from Whitehead's 'capital error' about the correlation of God and world. But Temple does go on at this point to deny once more that God and the world are correlative terms. He also strongly denies that the body/mind analogy, used by many Whiteheadian theologians, is at all adequate to throw light on the relation of the world and God. So, on the one hand, Temple rejects the analogy of the relation of an artist to his work on the ground that it makes God too external to the world; on the other hand, he rejects the analogy of soul to body on the ground that it too closely identifies God and world. The true analogy would seem to come somewhere between them, but what it might be is not said and is indeed hard to imagine.

The fundamental problem inherent in Temple's philosophy appears to me to be as follows. If one begins from the physical processes of a nature that is unfolding itself in space and time, and says that mind is an episode in the process, that it is secondary and derivative (and Temple says all these things), then I do not believe one can get beyond a strongly immanentist version of theism – indeed the process theologians and others call panentheism. Perhaps there is no need to get beyond this point, for panentheism may well be the form of belief that presents us with a conception of God that is not only the most intellectually coherent but also the most religiously satisfying and, in particular, one that accords with the Christian doctrines of incarnation and atonement. But Temple feels that he has to go on to a fully transcendent form of theism, though I do not think he can consistently reach this position from his starting point – not even by introducing what he calls a 'dialectical transition.' His dilemma is, in fact, much the same as that of his almost exact contemporary, Pierre Teilhard de Chardin. The two men were born in the same year, but of course Temple never got to know Teilhard's work, which was published only after both of them were dead. Clearly, the two had much in common. Teilhard's appreciation for the mysteries and potentialities of matter would have strongly appealed to Temple, and the interest that each of them had in evolution together with the optimistic spirit that this outlook tended to produce was another point in common. But they shared the same fundamental difficulty. When Teilhard announces at the beginning of *The Phenomenon of Man* that he proposes *only to see*, that is, to confine himself to phenomena,[17] he sets out on a path of immanence from which he eventually breaks out only by a *tour de force*. It may be noted, however, that both Teilhard and Whitehead employ an idea that considerably eases some of their problems – I mean the idea that everything that exists, even the most elementary particle or unit of energy, has a mental as well as a physical pole (Whitehead), an inside as well as an outside. This is the doctrine called 'panpsychism.' It did not commend itself to Temple, though it would cohere rather well with his doctrine of value, according to which the mind discovers kinship with itself among the things of nature.

I have said that Temple's philosophy flows naturally into his theology, and that one might expect this result if, as we have seen reason to believe, the philosophy is largely determined by Christian apologetic motives. But the theology is not left unaffected by the philosophy provided as its substrate. There is a reciprocity between the two, as we shall see.

Temple, according to his biographer, F.A. Iremonger, was never a Protestant, and his theology could be broadly described as a form of liberal catholicism. This fact again might be expected, because of his dialectical understanding of history. The catholic Christianity that developed into the medieval dogmatic system is the thesis. The Reformation is the antithesis. Here Temple draws an explicit comparison between the development in theology and the development in philosophy, and in particular between Luther and Descartes. The relation between these two men, he holds, is 'neither superficial nor accidental: both express one great principle, the principle of private judgment.'[18] But since, as we have seen, the synthesis is nearer to the thesis than to the antithesis, though it contains whatever valuable corrections find expression in the antithesis, modern theology will seek to preserve catholic truth in its integrity (and in fact Temple was a very orthodox theologian), but catholic truth as criticized and set free by the spirit that was born in the sixteenth century.

Two doctrines are especially prominent in Temple's theology: the doctrine of the incarnation and that of the sacraments. These emphases are very much in accord with the Catholic and Anglican tradition, but clearly they are very much in accord also with the tendency of Temple's philosophy and his stress on the reality of the physical and material. We may recall his often quoted assertion that Christianity 'is the most avowedly materialist of all the great religions ... Its own most central saying is, "The word became flesh," where the last term was, no doubt, chosen because of its specially materialistic associations.'[19]

Incarnation was a major concern of Temple throughout his career. 'The Divinity of Christ' was the title of a long essay he contributed in 1912 to the important symposium *Foundations*, subtitled 'A Statement of Christian Belief in Terms of Modern Thought.' This essay was followed in 1924 by the full-length study in Christology called *Christus Veritas*. His Gifford Lectures of ten years later argue that the knowledge of God to be derived from our natural knowledge of the world demands completion in an act of incarnation. I remarked a moment ago that Temple's concern with incarnation accords well with his philosophical interest in the material, but again we have to notice that the relation of philosophy and theology in his mind was a reciprocal one, so that his choice of a philosophy reflects his prior belief in incarnation. Already in *Christus Veritas* he was complaining that although idealism is a philosophy both theistic and spiritual, it cannot allow for a specific incarnation because it devalues concrete existence in the everyday world.[20] So again one might argue that his philosophical ideas were not so much an independent substructure from which he went on to theology as, rather, an interpretation of the natural world made in the light of a prior commitment to the Christian faith. His actual working out of incarnational doctrine is inspired more by St John's Gospel than by any philosophical system, though admittedly he sees St John, whom he called 'the most modern of theologians,'[21] as the teacher of a comprehensive, if mainly implicit, philosophy. No doubt he had to some extent been influenced by Edward Caird's profound respect for St John's Gospel.[22] But Temple himself declared 'that for as long as I can remember I have had more love for St. John's Gospel than for any other book.'[23] The depth of his insight into this gospel is amply attested by the fact that his *Readings in St. John's Gospel* is still in print more than fifty years after its first publication.

In his teaching on the sacraments, Temple again stresses the reality of matter and the respect that we ought to have for the material. Matter is not alien to us, as materialists who think of mind as a mere epiphenomenon may have held; but neither is matter unreal, as some idealists may have held. Matter is real, and when conjoined with spirit, can effect spiritual results. Temple takes quite an objective view of sacramental action and subscribes to a doctrine of *ex opere operato* – not, indeed, in any magical sense, but in the sense that because of the dignity of the material and its fitness to be a vehicle for the divine, the reality of the sacrament does not depend on the changing moods of those who celebrate it or even (though Temple seems less sure on this point) of those who receive it. But there are in fact many devotees of sacramental worship ready to testify that when their faith has been low, their religious experience minimal, or when they have been more sensible of God's absence than of his presence, regular sacramental participation has brought them through to a revival of faith and a renewed sense of God.

When we turn from Temple's theology to his social teaching, the transition is again a smooth one – we are moving within the ideas of a man who had consciously sought to unify his vision of life. One who makes the doctrine of incarnation central and believes that the very word of God has entered the material and historical sphere is bound to be concerned with the material and historical conditions of life. Temple is quite explicit too about the significance of the sacramental for one's practical attitudes. He writes: 'It is in the sacramental view of the universe, both of its material and of its spiritual elements, that there is given hope of making human both politics and economics, and of making effectual both faith and love.'[24] Of course, broadly the same motivation lay behind that whole tradition of Anglican social concern that found expression in the labours of priests in slum Tractarian parishes and in the Christian social thought of such men as Maurice Reckitt and Vigo Demant.

I have already mentioned how Temple from his early days was constantly engaged in activities aimed at humanizing and Christianizing the fabric of society. His little book *Christianity and Social Order*, published during wartime in 1942, has rightly become something of a classic and is much better known than his larger learned volumes on philosophy and theology. But he clearly states in it that the church's social policies must be firmly based on the doctrines that she proclaims. To someone reading that book almost half a century after its appearance, it may seem very mild and reformist. Yet it must not be forgotten that it had a considerable influence on the formation of post-war England. We should notice too that Temple was in some ways well ahead of his times. In those days the world's resources seemed unlimited and no one had heard of an environmental crisis. But we find Temple writing: 'Land is not a mere "material resource." The phrase "mother earth" stands for a deep truth about the relationship of man and nature.'[25] Of course, such teaching follows directly from the theory of value that he had propounded years before.

I said that Temple's social teaching might strike us today as mild and reformist. That was not meant to be an adverse criticism. A few years ago, I myself published a little book called *The Concept of Peace*. One reviewer said of that book that it was a reformist document in a revolutionary situation, and no doubt he meant this statement as a criticism, but I think that the Christian is bound to be reformist rather than revolutionary. Temple himself had a doctrine of original sin, and he points out that anyone holding such a doctrine must be intensely realistic and 'conspicuously free from utopianism.'[26] Like his contemporary Reinhold Niebuhr, he remains far from being superseded, and offers us a more realistic and, I believe, more Christian socio-political doctrine than do many of the present practitioners of 'political theology.'

Such then was the multifaceted yet unified Christian vision of this remarkable man. Some of us can still remember the high hopes that were entertained when he was elected archbishop of Canterbury in 1942. Alas, these hopes were never fulfilled. Ill health and the constraints of wartime prevented him from giving the leadership that was expected, and he died only a little more than two years after election. Moreover, the world by then was looking much bleaker than it had appeared when Temple was at the height of his powers. In 1942, he wrote in a letter: 'What we must completely get away from is the idea that the world as it now exists is a rational whole ... The world, as we see it, is strictly unintelligible.'[27] Was this statement a repudiation of his earlier beliefs? Certainly, it was a repudiation of an optimism that had occasionally found expression in his writings. It was not a repudiation of the vision in its essence, but he had come to recognize that the way would be longer and more difficult than he had once supposed. The operative words are 'at this time' and 'as we see it.' The hope that all things will be gathered up in God remains. For us, the world is even more fragmented than it was for him, but that is all the more reason why we should not lose his vision of a final wholeness.

NOTES

1 Edward Heath in his foreword to Temple's *Christianity and Social Order* (London: Shepheard-Walwyn 1976), 1 (hereafter CSO)
2 Edward Norman, *Church and Society in England* (Oxford: Clarendon Press 1976), 280
3 A.M.G. Stephenson, *Anglicanism and the Lambeth Conferences* (London: SPCK 1978), 161ff
4 Temple, *Nature, Man and God* (London: Macmillan 1934), 495 (hereafter NMG)
5 Owen C. Thomas, *William Temple's Philosophy of Religion* (London: SPCK 1961), 58
6 Ibid., 147
7 NMG, 57
8 Ibid.
9 NMG, 212

10 NMG, 261
11 NMG, 395
12 NMG, 265
13 NMG, 435
14 NMG, 480
15 NMG, 265
16 NMG, 269–70
17 Pierre Teilhard de Chardin, *The Phenomenon of Man* (London: Collins 1959), 35
18 NMG, 75
19 NMG, 478
20 Temple, *Christus Veritas* (London: Macmillan 1924), ix
21 Temple, 'The Divinity of Christ,' in *Foundations*, ed. B.H. Streeter (London: Macmillan 1920), 214
22 Edward Caird, *The Evolution of Religion*, 2 vols. (Glasgow: Maclehose 1893), 2: 217ff
23 Temple, *Readings in St. John's Gospel* (London: Macmillan 1939), vii
24 NMG, 486
25 CSO, 112
26 CSO, 61
27 F.A. Iremonger, *William Temple, Archbishop of Canterbury: Life and Letters* (London: Oxford University Press 1948), 537f

The Sacramental View of Reality: The Spirit–Matter Problem

D.R.G. OWEN

The relationship of spirit and matter is a recurring problem for human thought. In the medieval period of Western culture, sometimes described as the Age of Religion, spirit was predominant and matter tended to be denigrated. In the modern period, which may be called the Age of Science, the roles are reversed: matter is usually given priority and the reality of spirit is often called into question. The spirit–matter problem becomes part of the debate between science and religion.

Modern science, as it took shape in the seventeenth century, almost immediately came into conflict with the established religion of the West. The Christian faith was associated with a way of perceiving reality that had remained virtually unchanged for centuries. I call this the religious world-view. Science gave rise to an interpretation of reality that was entirely different. This was the new, revolutionary scientific world-view.

The old religious *Weltanschauung* was dualistic, other-worldly, and mythical. In the first place, it divided reality into two separate realms, the natural and the supernatural, the material and the spiritual, this world and the other world; the human being was similarly split into two parts, body and soul. Secondly, the other world was regarded as higher, more real, and in every way superior to this world. Life here below was thought of as a 'weary pilgrimage,' a 'vale of tears,' a temporary stage on the soul's journey to the higher realm.

The body was called the 'prison-house' of the soul and the purpose of religious exercises was to enable the soul eventually to shake off its physical fetters and return to its true home in heaven. Finally, unusual events in this world were explained by reference to supernatural agencies that belong to the other world but frequently intervene in this one to bring about miraculous events.

The modern scientific world-view is almost the exact opposite: it is anti-dualistic, this-worldly, and non-mythical. Science confines its interest to the space-time universe; it describes the world as a self-sufficient, unified system in which all events are interrelated in a coherent, intelligible network of natural cause and effect. The long warfare between science and religion that has characterized modern culture is really a conflict between these rival world-views.

There are two popular ways of resolving this conflict. The first involves a return to dualism: the spiritual realm is consigned to religion and the material to science. This convenient division of labour becomes unacceptable when its dualistic presupposition is no longer tenable. The second way of solving the spirit–matter problem is to deny the reality of the one in order to assert the reality of the other, as in philosophical materialism and metaphysical idealism. This Procrustean procedure must be rejected if justice is to be done to the claims of both kinds of reality.

William Temple proposed his sacramental view of reality as both a solution of the spirit–matter problem and a resolution of the debate between science and religion. He recognized that modern science makes it necessary to abandon the primitive and outmoded religious world-view.[1] He therefore took as his starting-point 'the picture which Science gives us.'[2] According to Temple, this picture shows us that reality consists of a series of grades – matter, life, mind, spirit – each of which depends for its existence on the grade below and for its completion on the grade above.[3] We can look at this series from two points of view. If we begin at the lower end we shall be tempted to explain everything in terms of matter and natural causes; this is the temptation of science. If we begin at the upper end we shall try to explain everything in terms of spirit and final causes; this is the temptation of religion. Temple maintains that the Christian religion is a special case since it has always insisted on the reality and importance of matter. Its central belief is that 'the Word was made flesh,' the last term being chosen 'because of its specially materialistic associations.' Christianity, said Temple in a famous phrase, is 'the most avowedly materialist of all the great religions.'[4]

Temple then proceeded[5] to expound his view of 'the sacramental universe' and of the way in which spirit and matter are closely interrelated. Just as I express my self, that is, my inner being or spirit, in all my words and acts, so also God expresses himself in the whole physical creation. And just as my words are signs and symbols conveying my meaning, so all natural facts and events are signs and symbols conveying the divine aims and purposes. They are sacramental in this sense, a sacrament being, in the traditional definition, 'an outward and visible sign of an inward and spiritual reality.' Matter in general is the 'symbolic instrument' of spirit. 'That is the formula which we suggest as an articulation of the essential relations of spirit and matter in the universe.'[6]

It is important to note here that Temple did not think of spirit as something extraneous, and added *ab extra*, to matter. On the contrary, taking his clue from A.N. Whitehead, he insisted that the development from matter to life and mind and spirit was a continuous, evolutionary process. When matter achieved a sufficient degree of complexity it became first conscious and then self-conscious. Consciousness is a product of matter organized in a special way: 'my consciousness is not something else within the entire organism ...; my consciousness is itself that organism, being not only physical but psycho-physical.'[7]

In the same passage, Temple pointed out that this conception of the intimate unity of spirit and matter is closely akin to Marxist materialism. Both views recognize that matter is chronologically prior to mind, that mind appears as part of the evolutionary process of matter, that mind as we know it never operates apart from the physical brain, but that mind is not reducible to matter. Against Marx, however, Temple insisted that once mind appeared it was bound to become dominant and capable of controlling the whole process.

The sacramental view of reality has obvious implications for the relationship of science and religion. The division of labour that assigns matter to science and spirit to religion becomes impossible. The fact is, as Temple said, that both science and religion are concerned with 'the entire field of human experience.'[8] This fact makes conflict inevitable but also points the way to reconciliation. Both science and religion are interested in the whole of reality, but from different points of view and with different purposes in mind. Science, because of its method and its aims, is limited to the quantitative and mechanical categories. Applied to human conduct, it will therefore explain everything in terms of natural causes, ignoring the role of purpose and the possibility of responsible action. Applied to religious beliefs, it will concentrate on their purely psychological origin, disregarding the possibility that they refer to an objective reality. In order to avoid conflict at these points it is only necessary that science, in its preoccupation with the 'lower categories,' should not deny the reality of the 'higher.'[9] For example, explanation in terms of natural and mechanical causes does not in itself preclude the operation of final causes or purposes: the preparation and reading of this paper involve a complicated set of physical events in my brain and nervous system, but these subserve my purpose which is, of course, the edification of my hearers.

On the other side of the picture, religion must accept the scientific account of the whole realm of the physical and the mechanical, even if this acceptance entails the abandonment of cherished convictions. According to Temple, reconciliation is really quite easy in principle. Science and religion both look at the whole of reality but for different reasons. Science 'seeks knowledge for the sake of understanding, while religion seeks knowledge for the sake of worship.'[10] Science studies nature in order to find out how it works; the Christian religion interprets the whole realm of matter as a symbol and sacrament of the divine spirit.

Scientists in general have great difficulty with the concept of 'spirit'; the very word makes them nervous; it sounds 'spooky.' I think the reason for this response is that they interpret the meaning of 'spirit' in the context of the old religious world-view with its dualism and supernaturalism. Temple's sacramental view suggests a quite different way of perceiving reality and a different understanding of 'spirit.' I propose now to develop these suggestions in my own way, with reference first to the human spirit and then to the divine spirit or God.[11]

In the case of the human realm, the problem is how to do justice to the scientific evidence without slipping into materialism and how to explain the nature of spirit without relapsing into dualism. I think the clue to a solution lies in the distinction between subject and object. Science is concerned with the objects that make up the external world. Because science was so successful in this sphere, there was a tendency to assume that everything, including human nature, could be explained in the same way. It is true, of course, that man is a part of nature and is therefore an object. But he is different from other objects in one fundamentally important respect: he is a subject as well as an object. Nature consists of things that are either inanimate objects like stones or living organisms like plants or psychosomatic organisms like animals. What differentiates the human being is not his possession of an immaterial soul but his status as a *self-conscious* psychosomatic organism. This is the meaning of the statement that he is a subject as well as an object.

Self-awareness is the distinguishing feature of human nature. Instead of simply registering the sense-perception 'tree,' a human thinks or says 'I see a tree.' He recognizes himself as a subject with everything else over against him as object – things, other people, the world, all encountered as objects of his awareness. In fact, it is only as he becomes aware of objects as such that he becomes aware of himself as subject.

Man's status as a subject does not liberate him from the limitations inherent in his status as an object. Like the other animals, he is a product of natural evolution and is conditioned by the material basis of his existence: human consciousness remains dependent on the physical nervous system and the brain; human behaviour continues to be largely shaped by physiological, psychological, and sociological factors. The physical and behavioural sciences deal with evidence of this kind and their findings must be taken into account in any comprehensive view of human nature. This scientific evidence often results, understandably enough, in materialistic and deterministic conclusions. But it is important to recognize that these conclusions are philosophical theories and not scientific hypotheses.

There is evidence of a different kind that can be cited against the theories of unqualified materialism and absolute determinism. This evidence has to do with man's status as a subject, as a self-conscious psychosomatic organism. It is this *status* that we refer to when we speak of the human spirit. The human spirit, in this sense, is to be understood not in terms of an alleged separate and independent soul but rather in terms of man's special type of consciousness. I have noted that Temple was prepared to accept the Marxist definition of human consciousness as simply a property of matter organized in a definite way. It emerges in the course of natural evolution; and evolution is itself a process in which the material ingredients are organized in ever new and more complex patterns. The emergence of self-consciousness enables the human being to detach himself from nature sufficiently to observe it, to ask questions about it, to predict its course, and so, within limits, to control it. Human nature is not entirely reducible to the rest of nature precisely because it is capable of knowing and controlling nature. It is science pre-eminently that exercises this kind of knowledge and power in our day. Science itself therefore is the strongest witness to the reality of the self-awareness that makes these achievements possible.

Temple pointed out that science, somewhat paradoxically, ignores the fact of self-awareness. It does so because it is preoccupied with objects and therefore abstracts from the subject who knows the objects. But from a philosophical point of view, as Temple said in a striking epigram, the fact of knowledge is more significant than all the known facts.[12] It is by virtue of his status as a knowing subject that man may be said to transcend nature: the knowing subject is not of the same order as the known objects. It is to this capacity and this status that we refer when we speak of the human spirit.

It is interesting to observe that almost all analyses of human nature from Plato to Freud have been trinitarian. I am therefore following an established tradition when I make my own threefold distinction. First, human existence has a strictly physical basis which is described by the physical sciences. Secondly, the human being is largely shaped by psychological and sociological conditions which, in so far as they can be quantified, are the concern of the behavioural sciences. Thirdly, all these factors are combined in such a way as to produce a special type of consciousness, namely self-consciousness. I suggest that this last-named capacity is what is meant by the term 'spiritual.' The first two categories have to do with man as an object, while the third indicates what it means to be a subject.

It would be easy at this point to slip back into the old division of labour that I have previously rejected. Just as it seems natural to assign the physical dimension to the physical sciences and the socio-psychological to the behavioural sciences, so it is tempting to assign the spiritual to religion. This division would be a mistake because the threefold distinction does *not* mean that there are three parts of human nature that can be treated separately. It is especially necessary to insist that the spiritual is not a *tertium quid* that can be cultivated in isolation but is simply the way in which all the other factors are organized.

When the behavioural sciences define man as a psychosomatic organism they know very well that the *soma* and the *psyche* are not separate entities: the human being is a unified organism in which the physical and the psychical are distinguishable but not separable. Similarly, to introduce a third category is not to suggest that in addition to the *soma* and the *psyche* there is a third entity called the *pneuma*. It is simply to argue that man is not only a psychosomatic organism or object, but also a self-conscious psychosomatic organism – that is to say, a subject.

It may be helpful here to refer to what Gilbert Ryle has called a 'category mistake.'[13] In body–soul dualism, which Ryle cites as an example, the term 'soul' is used as though it belonged to the same logical category as the term 'body,' the soul is regarded as the same kind of thing as the body, only more refined, a ghostly counterpart of the body. This usage is comparable to supposing that the term 'team' belongs to the same category as the term 'players,' as if, after inspecting all the players, one might then ask, 'But where is the team?' The team, of course, is not something that exists in addition to the players; it *is* the players organized in a certain way. Similarly, according to my argument, the human spirit is not something over and above the physical and socio-psychological constituents but simply *is* these constituents arranged in a definite pattern. It would, however, be equally mistaken to conclude that because the team is nothing apart from its members, it is therefore nothing at all. A good team is always more than a collection of individuals; we speak, significantly, of 'team spirit' as a highly important factor. It would be the same kind of mistake to suppose that because the spirit is nothing apart from its material constituents it is therefore nothing at all.

The way in which I view the relationship between spirit and matter should now be clear. They are obviously not two different substances as in the old body–soul dualism. At the same time, the one is not reducible to the other, as they are in philosophical materialism and metaphysical idealism. The term 'matter' covers all the quantitative factors while the term 'spirit' refers to the way in which these factors are combined, and combined in such a way that self-awareness emerges as the distinctive human capacity. The human being is a subject as well as an object. Further, the self as subject expresses itself in and through its psychosomatic organism. In this sense, matter is the vehicle for the self-expression of spirit: the outward and visible is the sign of the inward and spiritual. Thus we may say that the relationship of spirit and matter on the human level is sacramental.

I conclude with a brief note on the implications of this conception of the human spirit for our understanding of the divine spirit or God.

The popular Christian version of the old religious world-view includes an idea of God that is the theological counterpart of its body–soul dualism. Just as the latter describes the soul as an entity independent of the body, so the former conceives of God as a supernatural being entirely separate from, and over and above, the whole material realm. I want to propose an alternative approach to the question about God and his relationship to the world.

This approach is based on three analogies between man and God. I have maintained that on the human level the term 'spirit' does not refer to some alleged immaterial part of man but simply signifies his special status as a subject, that is to say his distinctive capacity for self-awareness which is also the source of his knowledge of, and control over, nature. In this sense man transcends nature. I want to argue, by analogy, that God's transcendence does not imply any kind of crude supernaturalism that pictures God as a separate supreme being who rules this world from the other world. Just as in the case of man, so also with God transcendence signifies not separation but self-awareness and it is God's self-awareness that is the source of his knowledge and his power.

This kind of transcendence is of course perfectly consistent with immanence. However (and this is my second analogy), just as man's transcendence over nature means that he cannot be entirely reduced to the level of nature, as materialism supposes, so God's transcendence means that he cannot be simply identified with the world, as pantheism believes.

My third analogy brings me back, by way of conclusion, to the sacramental view of reality. Just as the human spirit expresses itself in and through the psychosomatic organism, so the divine spirit expresses itself in and through nature and history. The world is the self-expression of God – the outward and visible sign of the inward and spiritual reality. The relationship between God and the world is sacramental. This sacramental solution of the spirit–matter problem avoids, first, the dualism and supernaturalism of the old religious world-view; secondly, the monism and pantheism of more sophisticated religions; and thirdly, the materialism and reductionism that the scientific world-view is often thought to entail. The sacramental interpretation tries to do justice both to the actual scientific evidence and to authentic religious experience.

It may be that this proposed solution of the ancient problem will only succeed in offending both the materialists and the supernaturalists, both the scientists and the theologians. But that outcome may be unavoidable.

NOTES

1 William Temple, *Nature, Man and God* (London: Macmillan and Co. 1935), 21–2
2 Ibid., 198
3 Ibid., 474
4 Ibid., 478
5 Ibid., 479ff
6 Ibid., 492
7 Ibid., 487
8 Ibid., 31
9 Ibid., 52
10 Ibid., 30
11 On what follows, see my *Body and Soul* (Philadelphia: Westminster Press 1956), ch. 7.
12 Temple, *Nature, Man and God*, 129
13 Gilbert Ryle, *The Concept of Mind* (London: Hutchison's University Library 1950), passim

The New Biology and *Nature, Man and God*

A.R. PEACOCKE

INTRODUCTION

'Nature, Man and God'

'If I were asked what was the most disastrous moment in the history of Europe I should be strongly tempted to answer that it was that period of leisure when René Descartes, having no claims to meet, remained for a whole day "shut up alone in a stove."'[1]

This passage, with its indirect repudiation of any dichotomy between the physical and the mental and spiritual, between the empirical and the rational (and surely one of the most arresting openings of any chapter of philosophical reflection), was written, not by any logical positivist, evolutionary naturalist, or body–mind identitist, but by the one whose seminal thinking on nature, man, and God we are honouring in this volume in the way he would most have wished. For, in 1939, seeing 'the theological scene to be changing in ways which his generation had not foreseen' (as another archbishop, Michael Ramsey, put it[2]) and looking into the future, already overshadowed by the Second World War, he wrote: 'We must dig the foundations deeper than we did in pre-war years, or in the inter-war years when we developed our post-war thoughts. And we must be content with less imposing structures. One day theology will take up again its larger and serener task, and offer to a new Christendom its Christian map of life, its Christocentric metaphysic. But that day can barely dawn while any who are now already concerned with theology are still alive.'[3]

Temple was well aware that the philosophy in which he had been trained – that blend of Platonism and Hegelian idealism which he had caught in his youth from the teacher Edward Caird to whose memory his 1932–4 Gifford Lectures, *Nature, Man and God*, were dedicated – was intellectually inadequate to provide the safe path that would lead mankind from its available knowledge to God's revelation of himself in Jesus the Christ through the Holy Spirit, and was seemingly irrelevant to the cataclysmic spiritual disasters that were overtaking the Western civilization that had nurtured this philosophy, in both Germany and Britain.

William Temple's quest to relate the Christian faith to contemporary philosophy and to show Christ as the key to the unity and rationality of the world had failed – and, wrote Michael Ramsey, 'nothing in his last years befitted his greatness more than the humility with which he acknowledged that his quest had failed, and that other tasks were superseding it.'[4] Temple was right, for the spiritual crises of the Western world made Christian thinkers especially responsive to Karl Barth and his majestic emphasis on letting God be God and letting his Word speak and transform mankind.

But today, more than three decades after the end of the war within which Temple then wrote and during which he died, and with a waning of the intellectual excitement generated by Barth in at least theological circles (though his general impact on the intellectual life of the West seems to have been slight), we can see that the quest that Temple so wistfully saw being temporarily set aside is as vital and necessary as ever, as he rightly and instinctively assessed in the whole set of his life. Barthianism (if not the massive contribution of Barth himself) has waned because at the end of the day the neo-Barthian retreat into the pure Word of God available through the Scriptures is not a viable route. It is indeed circular, for the further it takes us away from the presumed fallibilities of our natural minds to the supposed divine word in Scripture, the closer it brings us to the question 'How can we know that these scriptures, this tradition, is transmitting to us the genuine Word of God?' And there is no answer to such questions without resort to empirical inquiries into the nature of the biblical literature and of religious experience, as well as more philosophical inquiries into what we mean by such questions and the terms they contain.

A recent rereading of *Nature, Man and God* has reminded me of William Temple's percipience in detecting those broad features in the new knowledge of his day – mostly from the sciences – that gave a new perspective on the world of a kind that theology could ignore only at peril to its fundamental claims to speak truth and to see it whole. In the manner of his philosophical tradition, Temple continually ontologizes Matter, Mind, and Spirit, denoting them thus with capital letters, rather than referring to them, as we might, as processes and activities discernible in the world. But, we nevertheless find him taking the measure of the scientific world-view and discerning what it implies – and thereby raising for us here and now similar questions about the world-view of the sciences of our own day. For he perceived very clearly the need first to discriminate, then to relate, the various levels of analysis available in the sciences of his day and this, I would urge, is still a necessary task in any attempt by theologians to take seriously the knowledge afforded by the sciences. Since it is not my intention to mount a full reappraisal of *Nature, Man and God* and Temple's other relevant writings, let me quote one of the passages written by him on this theme:

Broadly speaking, the modern scientific view affords an apprehension of the world as existing in a series of strata, of such a sort that the lower is necessary to the actuality of the higher but only finds its own fullness of being when thus used by the higher as its means of self-actualisation. Without the mechanical basis in matter, there could be no life of the kind that we know. Without living matter – bodily organisms – mind, as we know it does not arise. Without animal mind (seeking means to an end presented as good) there could be no spirit such as we know (choosing between ends by reference to an ideal standard of good). Now such a scheme can be regarded from two points of view; but whichever is adopted, care must be taken to avoid obliterating what is evident from the other. We may begin at the lower end of the series, and then there is no doubt about the reality of the material world. But the fact that this is real, and is the necessary basis of the world of life and mind and spirit as known to us, must not lead us to the supposition that there is nothing in these which is not observable in the material world as such. In the last resort it is, no doubt, true that there is only one world, and each department in isolation is an abstraction. We can say if we like that there is only one substance, and that the different sciences study not different substances, but different modes of action and reaction on the part of the same substance. But then we must be careful not to say that, because the actions and reactions studied in physics and chemistry are certainly real, therefore those studied in biology, in aesthetics, in ethics, in theology, are either unreal or else are only complicated forms of the other group.[5]

This passage contains *in nuce* the foundation of Temple's reflection on the creative work of God within nature, the immanence of the transcendent, of his understanding of man as part of the natural created order and yet manifesting peculiarly the transcendence of the immanent, and (in the tradition of J.R. Illingworth[6]) of his insight into the Incarnation as the supreme revelation of God who is the Transcendent who is immanent, and whose immanence is, must be, transcendent.

The last sentence of Temple just quoted, with its warning against attributing reality only to the simpler levels analysed by the atomic and molecular sciences, is echoed in other passages in his work:

Science, in following its method of using the 'lowest' category applicable, is not entitled to deny the applicability of 'higher' categories but is only seeing how far it can go without them. Even if it can cover all the facts and hold them together by means of 'lower,' as for instance mechanical categories, it does not necessarily follow that the 'higher' categories, such as purpose, have no rightful application at all ... The positive work of Science, in giving an account of observable facts by its own method, never justifies Science in proceeding to negative inferences concerning other methods of interpretation, provided that these in their turn do not exclude the method of Science.[7]

This cautionary word seems to me to be as necessary today as it was when Temple first wrote it and indeed to be even more pertinent when we come to reflect on the implications of modern biological science for any system of thought that is believable today concerning nature, man, and God. For Temple, more explicitly than any other theologian that I know, was concerned to stress the 'materialism' of Christianity in its understanding of 'matter' as a real, existent entity and as a necessary vehicle of 'Spirit.'[8] And so the scientific account of matter and of living matter must always be incorporated into the total theological understanding.

As is well known, Temple in his penultimate Gifford Lecture (XIX) developed the Christian understanding of sacraments both as material instruments of God's action in effecting his purposes and as symbols of God's self-expression, and so a mode of his revelation; and wove these two strands together into his concept of the universe as 'sacramental.'[9] It is not my purpose to develop this penetrating insight further here, but to stress that that insight itself stems from Temple's own understanding of the hierarchy of levels that exist in the world and his recognition that there has been a development in time, with new levels appearing, dependent on and yet genuinely emergent from, the preceding, simpler levels.[10]

It is this relation between the hierarchical levels in natural systems and the related hierarchy of the sciences that has been the focus of the intensive discussion of 'reductionism' in the sciences. In my view, one's assessment of this problem – and so of the relations between the sciences – is crucial for any philosophically judicious assessment of the relation of scientific knowledge about the natural world to theological affirmations, especially when that scientific knowledge is concerned with the phenomenon of life.

Reduction in the sciences

Before embarking on an exposition of my main theme – of the implications for any theology of nature, man, and God of the 'new' biology – let me briefly state my own understanding of the relationship between the hierarchy of *levels*, or *systems*, actually present in the natural world and the hierarchy of the *sciences* that study them, in the light of the discussions of the last four decades on 'reduction' in the sciences.[11] Reduction is the process which is broadly being urged upon us when we are told that '*X* is nothing but *Y*' where *X* = biology and *Y* = physics and chemistry; or *X* = sociology and *Y* = biology; or *X* = psychology and *Y* = neurophysiology; or *X* = religion and *Y* = ideological function in society. Hence, the colloquial name of 'nothing buttery.' Analysis of what is being affirmed in such assertions turns out to be more complex than first appears, as witness the vast literature on the subject. In my view it is necessary to distinguish carefully (1) the hierarchy of the levels of natural *systems* from the hierarchy of *theories* about the systems, each usually characteristic of a particular named science; and (2) *methodological reduction* from compositionist approaches. Methodological reduction is the largely uncontroversial procedure in research, whereby complex entities are broken down into smaller units (which have to be discovered) and the relationships between these units studied 'from the bottom up,' as it were. In a 'compositionist' (or holistic) methodology the whole is examined, 'from the top down,' in its total activity and its functioning characteristics as a whole are investigated.[12]

Controversy can ensue in relation to *ontological* reduction – assertions about what complex entities actually 'are.' One form of ontological reduction is not controversial – clearly the laws (for instance, of physics and chemistry) of the simpler entities (for example, atoms and molecules) are still applicable to those same entities when they function in larger wholes (for instance, atoms and molecules in biological organisms). But controversy arises when reductionist assertions are made of the *X/Y* variety already instanced. Thus, in the case of biological systems, all would agree that the constituent atoms and molecules obey the laws of physics and chemistry, but is it *therefore* the ultimate aim of biology 'to explain *all* biology in terms of physics and chemistry' as F.H.C. Crick asserted?[13]

In analysing this question I have found it useful to distinguish the *processes* going on at the various levels being analysed in a hierarchical system, such as a living organism, and the *theories* about (and language about and concepts applicable to) the same levels.[14] A justified reductionism recognizes that no processes, at any level, are autonomous; that is, that complex processes at one level are the joint operations of processes analysable and describable at a lower level. But such non-autonomy (= reduction) of processes does not thereby imply reduction of the *theories* about them. At 'higher' levels of complexity some genuinely new features and activities emerge that require distinctive theories, language, and concepts to describe them, and these theories, and so forth, are autonomous – that is, they cannot be translated into the terms of theories applicable to the lower levels of analysis.[15] So one can be anti-reductionist in this epistemological sense (that is, advocating that some, at least, higher level theories are autonomous) without thinking there is 'anything else there' in the higher complex whole. For such an anti-reductionist, all the *processes* are non-autonomous (reducible) and there is no *thing* extra in the more complex form. So in relation to biology, one can be anti-reductionist and yet not vitalist.

It has subsequently been argued that this kind of anti-reductionist affirmation (of theory autonomy, as I would call it) itself implies an affirmation of the *reality* of what the terms of higher level theories refer to.[16] That is, one must accept a certain 'robustness' (Wimsatt[16]) of the whole entity and resist any attempts to dilute the reality of affirmations about it in favour of a supposed lower level of 'reality' (which, if it is that of atomic and subatomic particles, turns out anyway to have its own kind of elusiveness, as any particle physicist will avow).

With this background discussion in mind let us now turn to our main theme.

THE 'NEW BIOLOGY'

In the following I shall review briefly, but I hope not too inaccurately, some of those developments in modern biology which together have altered the style of the study sufficiently to cause it to be denoted, as the organizers of this conference have done, the 'new biology.' I have just mentioned that one aspect of the reductionist/anti-reductionist discussion is focused on purely methodological issues and I affirmed that this aspect did not cause much controversy among biological scientists. As Dobzhansky argued, both methodologies are equally necessary and complementary and each is incomplete without the other.[17] This distinction within the methodological approaches of the two styles of biological investigation provides a convenient binary classification of the 'new biology,' without, in itself, prejudging any general, philosophical (or theological) conclusions about the implications of the biology itself.

Methodologically 'compositionist' biology

Evolution

To be clear about this important theory, I cannot do better than preface the discussion with a quotation from one of the best recent expositions of biological ideas, that of the French Nobel prize winner François Jacob:

> There are many generalizations in biology, but precious few theories. Among these, the theory of evolution is by far the most important, because it draws together from the most varied sources a mass of observations which would otherwise have remained isolated; it unites all the disciplines concerned with living beings; it establishes order among the extraordinary variety of organisms and closely binds them to the rest of the earth; in short, it provides a causal explanation of the living world and its heterogeneity. The theory of evolution may be summed up essentially in two propositions. First, that all organisms, past, present or future, descend from one or several rare living systems which arose spontaneously. Second, that species are derived from one another by natural selection of the best procreators.[18]

Controversies today. It is hardly necessary to remind those of you who live on the western side of the Atlantic of the eruption of a phenomenon called 'creationism,' that is, of the view that species suddenly come into existence and have not evolved, by whatever natural processes, from other species. The new development in this controversy is that creationists are claiming their views as 'scientific' and not simply as religious and on this basis are beginning to claim a space in biological textbooks and teaching in schools and colleges in the United States. The motivation of this campaign is clearly ideological, linked as it is with the emergence of the so-called moral majority with its mix of ethical, political, and religious conservatisms. The existence of this phenomenon, which is far from being confined to the United States, for there is an active pressure group of this kind in Great Britain, does serve to exemplify the vast confusion that exists both about modern evolutionary ideas in biology – the evidence for evolution, the status of the theory of evolution as such and of ideas about its means and mechanism – and about the relation of biological evolution to theistic doctrines of creation. This fog of confusion permeates quite exalted levels. For example, the current exhibition on evolution at the British Museum (Natural History) – or as most of us have known it for most of our lives, the Natural History Museum of South Kensington – is prefaced by an excellent introductory display which raises the question of biological evolution as an *alternative* to the biblical doctrine of creation of Genesis. Nothing more, quite properly, is said on this matter in the displays that follow but it is intriguing to see a scientific exposition even hinting that it has fallen into the tactical trap of entering the debate on the terms of the creationists.

While the claim for 'creation-science' to be given equal time with the theory of evolution in the public school class-rooms of some half a dozen states in the United States is being settled by the law courts, with the American Civil Liberties Union one of the chief antagonists of the creationists (not to mention all leading biologists and theologians and clergy of the principal Christian denominations), the *status* of the theory of evolution has also become an issue in the pages of the scientific journal *Nature*.[19] The controversy there was between the British Museum (Natural History) and the writer of an editorial in *Nature* who attacked the staff of the former for including the phrase

'If the theory of evolution is true' in one of their recent exhibition brochures, on the grounds that the theory of evolution could scarcely still be an 'open question among serious biologists.'

The water of this particular dispute – about the British Museum exhibition and accompanying literature – is further muddied by a more technical dispute about 'cladism' (or 'phylogenetic systematics') which is a new method of classification (or rather a revised version of that of Hennig, 1966) that is in direct conflict with the older school of 'evolutionary systematics' of Simpson and Mayr, Cain, and others.[20] This dispute might seem to be an example only of esoteric in-fighting among biological taxonomists but it has nevertheless provoked accusations of crypto-Marxism against the supporters of the, admittedly disputed, cladistic classification. So although cladistics is strictly 'not about evolution, but about the pattern of character distribution in organisms, or the recognition and characterization of groups'[21] (that is, cladistics is a method of systematics and is not concerned with process), nevertheless debate concerning it readily acquires ideological overtones.

Similar ideological accusations have been generated in another current evolutionary controversy, this time not in the pages of *Nature*, but one that erupted at a conference on 'Macroevolution' held in Chicago's Field Museum of Natural History in October 1980. This controversy, or rather group of controversies, is concerned with the tempo of, mode of, and constraints upon evolution.[22] First Eldredge and Gould, and now others, have postulated that evolution with the branching of a new species from an established one – 'speciation' – occurs relatively rapidly over short periods, not well or easily represented in the fossil record, rather than by the gradual accumulation of the small changes that Darwin regarded as constituting the differentiation of species – small, slow changes that are well represented in the fossil record, but are regarded by Eldredge, Gould, and others as inadequate in themselves to represent speciation.[23]

Evolution is regarded by these biologists as 'punctuated equilibrium' – long periods (of the order of five to ten million years) of biological equilibrium, virtually 'stasis,' characterized by minor adaptations to a slowly changing environment, punctuated by relatively short bursts of rapid change (over periods of about fifty thousand years) during which speciation occurs. *How* such rapid evolution ('macroevolution') might occur in, for example, small populations isolated either geographically or reproductively, and what constrains the range of biological blueprints actually existing, also gave rise to controversy at this notable conference. We need not go into the details of these controversies now – for example, the fascinating question of whether 'hopeful monsters' are the clue to speciation; that is, members of an established species that have undergone a mutation (presumably in a gene controlling a crucial developmental pathway) so as to cause a markedly idiosyncratic morphology or metabolism of the individual, but a change that did not impair its fitness to reproduce. On this postulate, the mutated gene of such a 'hopeful monster' would spread through the population, even if it had no extra survival value at the time – but was utilized and manifested as favourable only under some later changed environmental conditions or habitat (itself, of course, dependent on the creature's *behaviour*, as constantly emphasized by Sir Alistair Hardy[24]). Again, the purely scientific issues have been mixed with ideological ones when, for example, one critic of 'punctuated equilibrium' expressed the fear that Marxists, who espouse major political changes through rapid and abrupt leaps from one form of social and political organization to another, might 'be able to claim the theoretical basis of their approach was supported by scientific evidence' for punctuated equilibrium, if this idea were established.[25]

The subtleties of how evolution occurs have long been known to biologists who accept the neo-Darwinist 'modern synthesis' – for example, the intricacies of the interplay between genetic constitution and environment via the phenotype (involving the possibility that the predisposition to adapt to the environment is itself under genetic control[26]) and the way in which the behaviour (and so the consciousness?) of some birds and mammals is itself a factor in altering their 'environment' and so affects what genotypes are selected. But even those who have long recognized such subtleties (and Professor J.

Maynard Smith protested that many of the ideas presented as new at the Chicago 'Macroevolution' conference were, in fact, part of the 'modern synthesis' of the last three or so decades) must have been somewhat startled by the proposal of Gorczynski and Steele that an acquired characteristic (tolerance to a specific antigen) can be inherited in mice. The Lamarckian thesis, widely thought to have been finally discredited, was thereby resuscitated. The experimental evidence is disputed and what is happening in this case is still unsettled.[27]

Another controversy in evolution theory that has been going on for some time is the role of so-called neutral mutations that have no selective advantage or disadvantage. Whether or not such mutations will survive into succeeding generations depends on chance fluctuations. The chance of perpetuation by chance fluctuation is greater in small populations; such fixation is called 'genetic drift' and many apparently selectively neutral features of organisms (for instance, arrangement of bristles in *Drosophila*, the 169 recorded variants of human haemoglobin amino acid sequences, butterfly wing details, human fingerprint patterns) are attributed to it by many biologists. However, the idea is still disputed partly because of the difficulty of mounting experimental tests of finely graded differences in fertility.

These controversies concern aspects of evolutionary theory whose issues can be identified, but there are many areas where our ignorance is such that their baffling nature reduces most biologists to relative silence.[28] The whole area of morphogenesis is an example, whether in its epigenetic (appearance of new structures), or its regulatory (embryonic) or regenerative (restoral of damaged structures) aspects – and the genesis of the organization of nervous systems, especially brains, is the most baffling area of all. The last mentioned is connected, of course, with the daunting problem of behaviour (instinct, behavioural regulation, and new patterns of behaviour inexplicable in terms of preceding causes). The approach to such problems varies from the determinedly reductionistic, such as F.H.C. Crick's 'the ultimate aim of the modern movement in biology is in fact to explain all biology in terms of physics and chemistry'[29] (however much new conceptual frameworks have to be provided to incorporate the physico-chemical accounts of highly complex systems), to the avowedly holistic, such as that of Sheldrake[30] who, repudiating even the attempts of an organismic biology to begin to explain the problem of morphogenetics and behaviour just indicated, resorts to a hypothesis of 'formative causation.' In his view, specific morphogenetic fields are responsible for the characteristic form and organization of systems at all levels (not only biological, but also physico-chemical) and these morphogenetic fields are derived from those of previous similar systems; that is, past systems influence the form of present systems acting across space and time. This latter, very unconventional theory, is regarded by the author as scientific because he claims it can be tested. Be that as it may, the fact that a biologist has had to resort to such a hypothesis at least emphasizes the radical and fundamental nature of many problems still facing biologists.

I mention these controversies and proposals as examples of the openness of discussion and exploration in the study of evolution in biology. Such discussion and openness is a proper characteristic of any active science, but is frequently misunderstood by the popular, and more reprehensibly the intellectually sophisticated, media to represent, in this case, another attack on the status of the whole theory of evolution and of Darwin's explanation of the basic mechanism by which it occurred.

The common origin and evolutionary relationships between all living organisms. Darwin's proposal of 'natural selection' as the natural process by which evolution, that is, the transformation of species (or 'speciation'), occurred had the impact it did in the history of biology because – although the idea of the evolutionary interconnectedness of living organisms had been widely courted, on the basis of the geological record and morphological similarities, since the time of Lamarck (1744–1829) and of Darwin's own grandfather, Erasmus Darwin (1731–1802) – the idea had seemed implausible without a reasonable *natural* mechanism for its occurrence, and that is what Darwin provided.

The proposition that living, and extinct, species of all living organisms are connected by evolutionary relationships became surprisingly and rapidly accepted by biologists as being the only broad concept that made any sense of the burgeoning data of biological observation, even though the acceptability of natural selection as its means had many ups and downs (not least in Darwin's own mind!) until by the end of the nineteenth century the laws of heredity were established by Mendel and, belatedly, became widely promulgated. The advent of the science of genetics, as a quantitative, statistical analysis of hereditary transmission, based on the Mendelian concept of discrete hereditary factors ('genes') controlling such transmission, combined with cytological and genetic investigation of the role of the chromosomes in the cell nucleus in hereditary transmission, led to the so-called modern synthesis (Julian Huxley's denotation in 1942) of Darwinian natural selection with Mendelian genetics as the basic explanation of evolutionary change.

I have outlined some of the current controversies about the rate and mechanism of evolution which may modify, for example, Darwin's own belief that evolution occurred by the gradual, slow accumulation of change (a view that made his theory statistically vulnerable until the Mendelian laws of heredity had been established). Darwin regarded as 'natural' the selection effected by the environment operating on the spectrum of individual differences present in any biological population of a given species, in a way analogous to the human selection of 'sports' among pigeons, dogs, or cattle to utilize them for breeding. These changes he regarded as random with respect to the needs of the organism in relation to its environment (its physical, nutritional, and biological ambience) but he saw the environment as selecting them 'naturally' by allowing some organisms with particular characteristics to reproduce more readily than others and thus establish themselves as predominant in subsequent populations. We now know that there are discrete hereditary controlling factors (the genes), that these undergo sudden, once-for-all, mutations (caused by various physico-chemical agencies), which are random with respect to the 'survivability' of the organism (that is, its ability to live long enough to reproduce). As we shall see, 'randomness,' of *this* kind, is now better established than in Darwin's own day, and even than in the 1940s when the 'modern synthesis' was proposed, through knowledge of the underlying molecular mechanisms that has come through *molecular* biology.

We shall have cause to discuss later the interpretation of this mechanism of evolution and its general implications. But, at this stage, let me stress that the proposition of evolution – that all forms of life, current and extinct, are interconnected through evolutionary relationships – is not in dispute among biologists. As a group of British Museum (Natural History) biologists said in a letter to *Nature* (defending, be it noted, their above-mentioned use of the phrase 'If the theory of evolution is true' for which a *Nature* editorial had taken them to task): 'What we do have is overwhelming circumstantial evidence in favour of it [the theory of evolution] and as yet no better alternative.'[31] By the nature of the case, the postulate of biological evolution cannot be falsified, in the sense of Karl Popper, by performing repeatable experiments whose outcomes are inconsistent with the postulate – nor can most theories of geology and of cosmology. These two sciences share with evolutionary biology a *historical* character, that is, they are concerned with reconstructing what *has* happened to living organisms, the earth, the solar system, the galaxies, and so on. Such theories cannot be subjected to either verifying or falsifying procedures of an experimental kind though they can nevertheless be judged by their consistency with present observations and with reasonable extrapolation of current conditions to the past, by their ability to make the most comprehensive sense of the widest range of observations, and by their own internal consistency. Until the advent of modern biochemistry and molecular biology, the observations with which the postulate of evolution had to be consistent were drawn mainly from comparative anatomy and the morphology of living and extinct organisms (the last mentioned confined to those with hard, skeletal remains), considered in relation to their locality and inferred climate and other environmental conditions.

But twentieth-century biochemistry has demonstrated many more fundamental similarities at the molecular level between all living organisms from bacteria to man – first, in many of their shared mechanisms for storing and using energy and for conveying substances across membranes, in the structures of their working proteins and structural polymers (both proteins and polysaccharides), and supremely in their genetic material. As François Jacob has put it: 'Over and above the diversity of forms and the variety of performances, all organisms use the same materials for carrying out similar reactions, as if the living world as a whole always used the same ingredients and the same recipes, originality being introduced only in the cooking and the seasoning.'[32]

Moreover, secondly, the phase of this science that called itself 'molecular biology' demonstrated not only that the prime carriers of hereditary information in all living organisms is nuclear DNA which, through RNA, transmits its 'information' to the sequencing of active proteins, but also that the code that translated the information from base sequences in DNA, via RNA, to amino acid sequences in proteins (and thence to their structure and function) was the *same* code in *all* living organisms. Now all searches for an explanation of why a particular triplet sequence of bases in the nucleic acid codes for the placing of a particular amino acid in a growing protein chain have failed to provide a completely unambiguous chemical basis for the same code to be operating in all present living organisms. The most widely accepted reason is that all living organisms have derived from an original, particular living conglomeration of matter that became self-reproducing (perhaps in the kind of way suggested by Eigen and his colleagues – see below) and just happened to have at that time (though not necessarily without *some* chemical basis) the coding relation now imprinted into all subsequent living organisms. In other words, the *actual* coding relation in all living organisms is quite arbitrary and is explicable and comprehensible only as being the particular coding relation that happened to be present in the conglomeration of living matter that first successfully reproduced itself enough to outgrow all other rivals (which may well at that time have had different genetic codes).

Thirdly, molecular biology has also provided another independent and powerful confirmation of the evolutionary relationships between many still living species through its ability to work out networks of evolutionary relationships by comparing the amino acid sequences in proteins with the same chemical function (for example, cytochrome C)[33] in widely different organisms. If these organisms are connected in some kind of evolutionary tree and if, as is known to be the case, the amino acid sequences of a given protein accumulate variations with time in the regions less critical to their function, then the more the same protein (defined by its function) differs in sequence in two species, the more the species are likely to be separated on any evolutionary tree.[34] In this way, and through a much more careful statistical analysis than I have been able to indicate here, 'trees' of relationships between living organisms have been able to be constructed entirely on the basis of *molecular* evidence. The striking fact is that these 'trees' entirely confirm (and often illuminatingly amplify) the network of evolutionary relationships previously deduced on morphological and palaeontological grounds by Darwin and his successor biologists. Like the previous evidence (from the genetic code) the molecular studies provide entirely independent evidence for divergent evolutionary relationships from common origins, indeed *a* common origin – as does, incidentally, the remarkable *constancy* of the amino acid sequences of the histone proteins in eukaryotic cells (the proteins that surround and, possibly control, the DNA in the chromosome of all such cells and so have the same function, in a protected milieu, in all eukaryotic organisms).

Thus, again and again, the evolutionary hypothesis (if that is what we still, neutrally, prefer to call it) has survived the test of consistency with observation of a kind (at a molecular level) unthinkable even four decades ago when the 'modern synthesis' first emerged. Again this fact does not preclude controversy about the tempo of, mode of, and constraints upon evolution, but it renders entirely reasonable our basing our philosophy and theology on what we can presume to be the 'fact' of biological evolution, a process within which man is entirely included with respect to the biological and molecular aspects of his existence relevant to his origins.

Trends in evolution. We are bound to ask if there are any trends discernible in the processes of cosmic development and, in particular, of biological evolution and whether any of them are particularly relevant to man. This is, of course, a notoriously loaded question which men are only too ready to answer on the basis of their own significance in the universe grounded on their own importance to themselves! Is there any objective, non-anthropocentrically biased, evidence for directions or, at least, trends in evolution? Biologists are cautious about postulating 'progress' in evolution for the criteria of progress are often already chosen with man's special exemplification of them in mind, deliberately or otherwise. As G.G. Simpson says, 'Within the framework of the evolutionary history of life there have been not one but many different sorts of progress.'[35] He instances the kinds of progress as: the tendency for living organisms to expand to fill all available spaces in the livable environments; a succession of dominant types in biological evolution; the successive invasion and development by organisms of new environmental and adaptive spheres; increasing specialization with its corollary of improvement and adaptability; increasing complexity; increase in the general energy or maintained level of vital processes; protected reproduction–care of the young; change in the direction in which increase in the range and variety of adjustments of the organism to its environment occurs; individualization.

Perhaps it is more rewarding to discern any trends in evolution that reach their maximum expression in man. Because he is innocent of any desire to support Christian theism it is worth quoting in full Simpson's judgment on mans' place in nature:

> Man has certain basic diagnostic features which set him off most sharply from any other animal and which have involved other developments not only increasing this sharp distinction but also making it an absolute difference in kind and not only a relative difference of degree. In the basic diagnosis of *Homo sapiens* the most important features are probably interrelated factors of intelligence, flexibility, individualisation, and socialisation. All four of these are features that occur rather widely in the animal kingdom as progressive developments ... In man all four are carried to a degree incomparably greater than in any other sort of animal ... In other

respects, too, man represents an unusual or unique degree and direction of progress in evolution.

He concludes, 'Even when viewed within the framework of the animal kingdom and judged by criteria of progress applicable to that kingdom as a whole and not peculiar to man, man is thus the highest animal.'[36]

There is one broad feature which seems to be common to both cosmic and biological development and, indeed, to human social and cultural history. It is the tendency for more and more complex structures to emerge in the world. Attempts to find a suitable measure of the complexity of structures by means of information theory have not been as useful as was earlier hoped.[37] The need for clarity here is provoked by the proposal that there is a connection between 'complexity' and consciousness. This correlation has been strongly urged and gained wide currency through the writings of Teilhard de Chardin, who calls this his 'law of complexity-consciousness.'[38] This 'law' is certainly an impression, though an imprecise one (what *kind* of complexity is to be correlated with consciousness?), that is given by the broad sweep of evolution, but Teilhard's unsubstantiated panpsychic assumptions give grounds for doubting such a sweeping generalization. Until we can quantify 'complexity' better it is unwise to promote our impressions into a 'law' that can then tempt us into applying it, for example, to the undoubted complexity of intra- and inter-communicating human societies.

Implications of biological evolution for theology. Let us now consider some consequences of biological evolution for any theology that has a doctrine of creation, namely, one that affirms that all-that-is is not self-explanatory and owes its very being to a transcendent Creator who, in principle, is other than the world that is created, however much it may be an expression of that Creator's fundamental inner being.

The process of evolution is continuous with that of inorganic and cosmic evolution – life has emerged as a form of living matter and develops by its own inherent laws, both physico-chemical and biological, to produce new forms with new emergent qualities requiring new modes of study (the various sciences) as well as new concepts and languages to describe and explicate them. So the stuff of the world has a continuous, inbuilt creativity – such that, whatever 'creation' is, it is not confined to a restricted period of time but is going on all the time (and indeed modern physics would support seeing time itself as an aspect of the created order). So, if we identify the creativity of the world with that of its Creator, we must emphasize that God is *semper Creator*, all the time creating – God's relation to the world is perennially and eternally that of Creator. But to speak thus is to recognize also that God *is creating* now and continuously in and through the inherent, inbuilt creativity of the natural order, both physical and biological – a creativity that is itself God in the process of creating. So we have to identify God's action with the processes themselves, as they are revealed by the physical and biological sciences, and this identification means we must stress more than ever before God's *immanence* in the world.

The processes of evolution, initially the physical and cosmological, and then more strikingly the biological, are characterized by the *emergence* of new forms within and by means of continuous developments subject to their own inherent, regular, law-like behaviour that is studied by the sciences. What emerges is usually more complex and, along certain branches of the evolutionary tree, more and more conscious, culminating in the self-consciousness and the sense of being a person that characterizes humanity. In theological terms, God's immanent creative action in the world generates within the created order a being, the human being, who becomes self-aware, morally responsible, and capable of himself being creative and of responding to God's presence. Thus the natural, biological, and human worlds are not just the stage of God's action – they *are* in themselves a mode of God in action, a mode that has traditionally been associated with the designation Holy Spirit, the Creator Spirit. I think that to give due weight to the evolutionary character of God's creative action requires a much stronger emphasis on God's immanent presence in, with, and under the very processes of the natural world from the 'hot, big, bang' to humanity.

The evolutionary account of mankind, as depicted in geology, palaeontology, biology, and anthropology, demonstrates unambiguously that man is a creature who has emerged into a self-consciousness that has enhanced his adaptational flexibility and power over his environment, and so his biological survival ability. The only state of primeval 'innocence' that such a scientific account of the emergence of man might allow is of a kind that can be attributed to all non-human, non-self-conscious mammalian organisms, namely, one innocent of any sense of responsible choice and of the relationship of power to moral choice. In the emergence of man, science sees only a gradual awakening to self-consciousness and so to an awareness of freedom of moral and other choice, and also to power over the environment. The traditional theological doctrine of 'the Fall' as a disobedient act by the original man and woman (Adam and Eve) in which they fell from grace and which so altered their state that they transmitted this state of 'original sin' pseudo-genetically to all succeeding mankind is clearly at odds with the scientific account – which sees in man only an emergence from the consciousness of the higher mammals to self-consciousness, language, and deliberate choice. As I do not need to remind you, the meaning of what is now regarded as the Genesis myth of 'the Fall' has been widely and profoundly interpreted existentially in our times by both biblical and systematic theologians as a myth of man's present *state* of alienation from God, and of disharmony both between men and between man and nature.[39] It should hardly need stressing that such a reappraisal of what has traditionally been called 'the Fall' of man should have a profound effect on how Christians ought to conceive of the innovatory and compensating 'redemption' wrought by Christ – it *should* hardly need stressing but for the fact that most Christian soteriology seems to wend its blissful way, whether as evangelical preaching or as systematic theology, in a state of even greater innocence than Adam himself of knowledge of the present state of understanding of *both* the evolution of man *and* the interpretation of the Genesis myths.[40]

Ecology

Interconnectedness. 'It is hard to be a reductionist ecologist' according to Dr Norman Moore, an eminent ecologist, and this increasingly important branch of biology – amazingly *un*fashionable even two decades ago – certainly qualifies as one for which a 'compositionist' approach is essential. For it is the study of the dependence of all living forms not only on the continued supply of energy from the sun and on the physical and natural environment, but also on each other. All plants and animals, including man, live in intricate systems consisting of many cross-flows and exchanges of energy and matter, of a labyrinthine complexity that has, until recently with the advent of computers, defied analysis. The impact of man on evolved, natural ecosystems has been dramatic, often catastrophic, and frequently disastrous – though not always, as in the (until recently) successful maintenance of the fertility of agricultural land in many parts of Europe over hundreds and, in some areas (for instance, northern Italy), thousands of years of careful husbandry. What we now consciously realize, which past generations had only intuitively sensed, is that through his technological power man is today forced into becoming a manager of the earth whether he likes it or not, for almost everything man does to natural ecosystems for his own presumed benefit has long-term global effects for better or for worse.

Implications of ecology for theology. This interconnectedness of all living systems on the surface of the earth with each other and their interaction with their physical environment is one expression of a more general unity and mutual interconnectedness and interdependence of all things and events. That the universe is such a complicated web of relations between the parts of a unified whole has, according to Fritjof Capra,[41] been a major emphasis in much Eastern mystical thought.

If in its understanding of the doctrine of creation Christian theism were to re-emphasize, as I have urged, its earlier stress, as in the *Logos* doctrine, on the *immanence* of God as Creator in the very creative processes themselves, it could bring a new dimension to bear on the debate in ecologically sensitive circles about how 'ecological values' are to be generated and sustained. For if the subtle, interlocking processes of the world's ecosystems are in themselves God in action as continuous Creator, then we must interact with them in a way that is so imbued with sensitivity and an intelligent apprehension of their complexity that it could properly be denoted as 'reverent' – that is, as exhibiting that 'reverence for life' that Albert Schweitzer used to urge upon us. Such an attitude combines a new theological 'input,'[42] which could motivate and maintain a sensitivity to the value of ecosystems that exist now, with a parallel recognition that we depend on the biological science of ecology. For through the results of its investigations ecology can provide us with the knowledge that makes possible any intelligent appraisal of our actions upon such ecosystems. A happy marriage of theology and science?

Methodologically 'reductionist' biology[43]

Molecular biology

Its history and present state. 'Molecular biology' is the name given to that phase in the development of biology in which, largely, physicists and chemists turned their attention to the molecular basis of fundamental biological processes. It is scarcely distinguishable from the science of biochemistry except for the attitudes of the pioneers of this phase of biological science, among whom at least one group was, oddly, antipathetic to biochemistry as it had developed during the first four decades of this century. This was what G.S. Stent has called the 'informational school' in a 1968 article reviewing the stage molecular biology had then reached, an article entitled, it must be noted, 'That Was the Molecular Biology, That Was.'[44] This 'informational school' consisted mainly of physicists who were, Stent affirms, 'motivated by the fantastic and wholly unconventional notion that biology might make significant contributions to physics.'[45] He traces back to Niels Bohr this motivation which was transmitted to his pupil Max Delbrück who in 1938 started his work on bacteriophages and so initiated what Stent calls 'the romantic phase' of molecular biology.[46] In this phase the informational school, which was largely German in origin and later based in the United States, thought that 'the *real* problem requiring explanation is the physical basis of genetic information' – and this phase was vastly stimulated in 1945 by the immensely influential *What Is Life?* written during the war by the famous quantum physicist Erwin Schrödinger, then in Ireland. The other principal strand in this phase of molecular biology, from about 1938 to 1952, was the 'structural school' which was mainly of British provenance (under the influence of W.T. Astbury, J.D. Bernal, and their pupils) and later also American (for example, Linus Pauling). Indeed W.T. Astbury was the one to coin the label 'molecular biology' for the study 'concerned with the *forms* of biological molecules and with the evolution, exploitation and ramifications of these forms in the ascent to higher and higher levels of organization. Molecular biology is predominantly three-dimensional and structural.'[47] In contrast to the 'informational' molecular biologists, members of this school were preoccupied with structure rather than information and 'reflected a down-to-earth view of the relation of physics to biology – namely, that all biological phenomena, no matter what their complexity, can ultimately be accounted for in terms of conventional physical laws.'[48]

These two schools of molecular biology converged personally in the meeting in 1951 in Cambridge, England, of the young American biologist J.D. Watson, seeking the basis of genetic information and its transmission by the hereditary mechanism, and the English physicist F.H.C. Crick, studying biological structures by X-ray diffraction. The story of how they came to postulate the double-helical *structure* of that material, DNA, which turned out amazingly to be the conveyer of genetic *information* by virtue of its structure has been told at least three times – once autobiographically (and not a little scurrilously) by J.D. Watson, in a more historical perspective by R. Olby, and in great detail, based on interviews with many of the participants, by H.F. Judson.[49] So there is no need for me to recall this fascinating history – of which I had a grandstand view as an investigator, at the time, of the solution properties of DNA. The publication of the structure of DNA by Watson and Crick in 1952 (in association with the X-ray diffraction studies, by M.H.F. Wilkins and others in London, on which it was based) led to a veritable explosion in 'molecular biology' and initiated a second 'dogmatic phase' (as Stent designates it). This phase of molecular biology, which lasted approximately until 1963, was dominated by the 'central dogma' of molecular genetics, of Watson and Crick, that DNA replicates its unique sequence of units (nucleotides) autocatalytically by copying one of its two intertwined chains and also acts as a template for single RNA chains, which then control the synthesis and amino acid sequences of proteins. The properties of proteins depend on these sequences, so

$$\text{DNA} \rightarrow \text{DNA} \rightarrow \text{RNA} \rightarrow \text{protein}$$

would be the sequence of transfer of information. The 'dogmatic phase' saw this proposal vindicated, enriched by Jacob and Monod's ideas on messenger RNA and the operons. So the molecular machinery that transmits the information came to be elucidated.[50]

If this phase represented one of Kuhn's paradigmatic shifts in the history of biology, the period since 1963 – the 'academic phase' (Stent) – sees molecular biology reintegrated with biochemistry and becoming a 'normal science,' with a huge increase in the number of its practitioners and range of study – and with medical and technological consequences of which we are hearing more and more. Those with a pioneering spirit that prefer to confront mysteries in the natural world rather than to solve the problems it poses are turning to the study of the genetics, development, and mode of operation of the higher nervous system and to morphogenesis. These are regarded by some as the last frontiers of biology, but past experience has shown only too often that such frontiers, like the false summits well known to the slogging mountaineer, almost always give way to yet others as the vista changes.

In all of this development no other laws of physics or chemistry had had to be propounded to unravel the molecular processes of genetic information storage and transfer, and to this extent the 'romantic' aspirations of the informational school had not been fulfilled – although, in fact, it was they as much as any who largely elucidated, during the 'dogmatic phase,' how phage DNA replicates itself and controls the synthesis of the specific proteins of such viruses. Most molecular biologists would now agree with Sydney Brenner when he wrote, in the 1974 issue of *Nature* surveying molecular biology, that 'much has been written about the philosophical consequences of molecular biology. I think it is now quite clear what the enterprise is about. We are looking at a rather special part of the physical universe which contains special mechanisms none of which conflict at all with the laws of physics. That there would be new laws of Nature to be found in biological systems was a misjudged view and that hope or fear has just vanished.'[51] No 'conflict at all with the laws of physics' – agreed, but does this judgment mean that all accounts of biological systems are to be subsumed into physics? Does the triumph of molecular biology really imply the long-term demise of all 'compositionist' or 'holistic' approaches and the final victory for a reductionist interpretation of biology in the terms of F.H.C. Crick as already quoted?

The philosophical interpretation of molecular biology

(a) *Reductionism, anti-reductionism, and vitalism.* Certainly Crick thought so and prefaced his 1966 *Of Molecules and Man* with a quotation which, judging by the subsequent contents, as indeed by Crick's well-known attitudes, leave us in no doubt that it is meant to be taken ironically. The quotation is from Salvador Dali: 'And now the announcement of Watson and Crick about DNA. This is for me the real proof of the existence of God.' The irony comes in because Crick identifies belief in God with vitalism – the view that living organisms have some special added entity or force over and beyond non-living matter – for, as Crick saw it, molecular biology had triumphantly demonstrated the *molecular* basis of the most distinctive feature of living organisms, their ability to reproduce, and thereby rendered all such vitalism null and void. This equation of vitalism with theism is, I think, simply false, though understandable because Christian apologists have unfortunately had, and still do have, a tendency to attempt to insert 'God' into the gaps of biological explanation.

Oddly, Stent thinks Dali is right but, in my view, for the wrong reason![52] for Stent identifies 'God' with the rationality of the universe and its amenability to rational explanation – 'God' is the single principle that regulates everything and makes science possible. Belief in 'God' so formulated is the axiom from which it follows that an explanation of the world is accessible to human reason. Stent thinks that the structure of DNA, and molecular biology as a whole, is further proof for such a 'God,' even when, as for Crick himself, this 'God' is identified with 'Nature,' without remainder. But Stent himself wishes to emphasize not the triumph of molecular biology over vitalism (which, incidentally, he agrees with me is *not* implied by theism), but rather 'the apparent inaccessibility of the human psyche to scientific study' and looks to Eastern philosophical-ethical systems (Confucianism and Taoism) as providing for man's harmony, both inner and outer, with his environment and offering the possibility of abandoning the Platonic 'God' of rationality.

Stent is unrepresentative among molecular biologists in taking this line and most, when they comment on philosophical or theological matters, follow Crick in finding reinforcement of their reductionist views from modern molecular biology. K.F. Schaffner seems to argue the reductionist case with respect to the non-autonomy of the *processes* going on at the higher level in relation to molecular processes, but goes on to recognize that a 'compositionist' (holistic) methodology is often the most pragmatically feasible, even allowing the formulation of 'specifically biological theories.'[53] He appears to me not to distinguish clearly enough between the reducibility (= non-autonomy) of *processes* in higher levels to lower levels, with which I would agree, from the autonomy (= irreducibility) of *theories* concerned with the lower levels (for instance, physics and chemistry). The discussion of these matters is often confused because of a failure to make this distinction of Morron Beckner, and one even finds Schaffner, who describes himself as an in-principle reductionist, using the same arguments for his position as Michael Polanyi, who regarded himself as being on the opposite side of the fence.[54]

But the climate has changed since the successes of the 'dogmatic phase' of the 1960s and one finds increasing recognition of the need

y new concepts to describe the complexities of living mat-
pts not even 'in principle' conceptually reducible to those of
ics and chemistry we have hitherto known. This is what I,
g Beckner, would regard as genuine 'theory autonomy,' that is, non-reducibility of higher level concepts; such views have been propounded in relation to both dissipative systems and the application of network thermodynamics to biological systems.[55]

In my view, the real irony is that the autonomy of at least some of the concepts applicable to the higher levels of complexity of biological systems is, in fact, implicit in the inbuilt dichotomy that we saw had characterized the pioneering days of molecular biology – namely, the different aims and directions of the 'informationists' and the 'structuralists.' The convergence and union, already mentioned, of these two movements in the information-carrying structure of DNA shows precisely what the non-reductionist is trying to argue. For in no way can the concept of 'information,' the *concept* of conveying a message, be articulated in terms of the *concepts* of physics and chemistry, even though the latter can now be shown to explain how the molecular machinery (DNA, RNA, and protein) operates to convey information. For the concept of 'information' is meaningless except with reference to the functioning of the whole cell, itself conceived in relation to its genetic and evolutionary history. In no textbook of chemistry or physics, *as such*, would (indeed, do) the concepts of 'information theory' have to be expounded to understand molecular chemistry or atomic physics. It is a concept applicable to the molecular system of DNA–RNA–protein only when these complex structures are linked in networks of interrelations constituting the whole cell. Thus (and *this* is the irony) the marriage of 'information' and 'structure' which characterizes molecular biology is a classic illustration of the kinds of distinction we need to make concerning the autonomy of *concepts*, *theories*, and so on (for instance, 'information') and the non-autonomy, that is, reducibility, of *processes* (there is nothing else going on in the replication of DNA, for example, except rearrangement of atoms and molecules). Thus the anti-reductionist position requires no mystical affirmations of vitalism, or any like 'non-natural' hypotheses, but it does require the recognition in many cases of the autonomy, and so of the validity, of the concepts and language we apply to the higher levels of complexity of the natural world, in this case the biological. There are simply not just grades of 'reality' of which atoms are the 'most real,' biological entities less so, and persons the least.[56]

(b) *Chance, law, and the origin of life.* Until the recent past, chance and law (necessity or determinism) have often been regarded as alternatives for interpreting the natural world. But the interplay between these principles is more subtle and complex than the simple dichotomies of the past would allow. In any particular state of a system, we have to weigh carefully the evidence about the respective roles of these two principles in interpreting its present and past behaviour. The late Jacques Monod contrasted the 'chance' processes that bring about mutations in the genetic material of an organism and the 'necessity' of their consequences in the law-abiding, well-ordered, replicative mechanisms which constitute that organism's continuity as a living form.[57] He pointed out that mutations in the genetic material, or DNA, are the results of physico-chemical events and that their locations in the molecular apparatus carrying the genetic information are entirely random with respect to the biological consequences to and needs of the organism. The two causal chains are entirely independent and so Monod is correct, to this extent, in saying that evolution depends on chance. This is the basis on which Monod stresses the role of chance: 'Pure chance, absolutely free but blind, at the very root of the stupendous edifice of evolution.'[58] There is, according to Monod, no general purpose in the universe and in the existence of life, and so none in the universe as a whole. It need not, it might not, have existed nor might man.

However, I see no reason why this randomness of molecular event in relation to biological consequence has to be raised to the level of a metaphysical principle interpreting the universe. For in the behaviour of matter on a larger scale many regularities that have been raised to the level of being describable as 'laws' arise from the combined effect of random microscopic events which constitute the macroscopic. So the involvement of chance at this level of mutation does not, of itself, preclude these events manifesting a lawlike behaviour at the level of populations of organisms and, indeed, of populations of biosystems that may be presumed to exist on the many planets throughout the universe that might support life.

Instead of being daunted by the role of chance in genetic mutations as the manifestation of irrationality in the universe, it would be more consistent with the observations to assert that the full gamut of the potentialities of living matter could only be explored through the agency of the rapid and frequent randomization which is possible at the molecular level of DNA. This role of chance is what one would expect if the universe were so constituted as to be able to explore all the potential forms of organizations of matter (both living and non-living) which it contains. I see no objection to conceiving of God as allowing the potentialities of his universe to be developed in all their ramifications through the operation of random events; indeed, in principle, this is the only way in which all potentialities might eventually, given enough time and space, be actualized. It is as if chance is the search radar of God, sweeping through all the possible targets available to its probing. Since Monod wrote his book there have been developments in theoretical biology over the last decade that cast new light on the interrelation of chance and law (or necessity) in the origin and development of life.

Ilya Prigogine and his colleagues at Brussels, who were already well known for their work on the thermodynamics of irreversible processes, have in recent years increasingly turned their attention to the analysis of living systems. They asked how it was that such highly ordered systems as living organisms could ever have come into existence in a world in which irreversible processes always tend to lead to an increase in entropy, in disorder. They have been able to show that there exists a class of open systems, 'dissipative structures,' that can maintain themselves in an ordered, steady state far from equilibrium. Under certain conditions they can undergo fluctuations that are no longer damped, as they are near to equilibrium, but are amplified so that the system changes its whole structure to a *new* ordered state in which it can again become steady and imbibe energy and matter from the outside and maintain its new structured form. It turns out that the conditions for such instability and transition are not so restrictive that no systems can ever possibly obey them. Indeed, systems such as those of the first living forms of matter, which must have involved complex networks of chemical reactions, are very likely to do so. Because of the discovery of these dissipative systems and of the possibility of 'order-through-fluctuations,' it is now possible to regard the emergence of highly ordered and articulated molecular structures that are living as highly probable on the basis of these physico-chemical considerations. To this extent, the emergence of life was inevitable but the form it was to take remained entirely open and unpredictable. Prigogine and Nicolis go further: 'We are led to a first parallelism between dissipative structure formation and certain features occurring in the early stages of biogenesis and the subsequent evolution to higher forms. The analogy would even become closer if the model we discuss has further critical points of unstable transitions. One would then obtain a hierarchy of dissipative structures each one enriched further by the information content of the previous ones through the 'memory' of the initial fluctuations which created them successively.'[59]

But how can a molecular population have 'information content,' and how can it store a 'memory'? It is to problems of this kind that Eigen and his colleagues at Göttingen have directed their attention. They have examined the changes in time of a population of biological macromolecules, each capable of carrying the information required to make a copy of itself (as can DNA). Their treatment is based on the theory of games and of time-dependent random processes, but they have been able also to illustrate the principles involved by inventing actual games that the novice can play (with, for example, octahedral dice!).[60] They have been able to delineate fairly precisely what kind of combination of chance and law will allow such a population of information-carrying macromolecules both to develop into one 'dominant species' and to yet have enough flexibility for further evolution.

Moreover, they have been able to propose what kind of self-organizing cycles of macromolecules would most likely be viable and self-reproducing. Eigen concludes thus: 'the evolution of life, if it is based on a derivable physical principle, must be considered an *inevitable* process despite its indeterminate course ... it is not only inevitable "in principle" but also sufficiently probable within a realistic span of time. It requires appropriate environmental conditions (which are not fulfilled everywhere) and their maintenance. These conditions have existed on Earth.'[61] According to this analysis, although the emergence of living systems may be 'inevitable,' it is nevertheless 'indeterminate.' For it is impossible to trace back the precise historical route or to predict the exact course of future development, beyond certain time limits, because of the involvement of time-dependent random processes.

The work of Prigogine and Eigen and their collaborators now shows how subtle can be the interplay of chance and law (or necessity), of randomness and determinism, in the processes that lead to the emergence of living structures. These studies demonstrate that the mutual interplay of chance and law is in fact creative within time, for it is the combination of the two that allows new forms to emerge and evolve. This interplay of chance and law appears now to be of a kind that makes it 'inevitable' both that living structures should emerge and that they should evolve – given the physical and chemical properties of the atomic and subatomic units in the universe we actually have. One obtains the impression that the universe has potentialities which are becoming actualized in time by the joint operation of chance and law, of random, time-dependent processes in a framework of lawlike determined properties – and that these potentialities include the possibility of biological, and so of human, life.

Implications of molecular biology for theology. I have tried to give a fair account of one aspect – the interplay of chance and law in the life-game – of our present scientific perspective in the world, and to draw out its general implications, so far without much reference to a theism that conceives of a *creator* God. We are inevitably involved in thinking again what the assertion that there is a God who is Creator really can mean in this new context. I suggest we need to develop some new models and analogies. For we now see the role, in the eliciting of life (and so of man), of the interplay of random chance micro-events with the necessity that arises from the stuff of this world having its particular 'given' properties. These potentialities are written into creation by the Creator himself and they are unveiled by chance exploring their gamut, a musical term which has come to mean 'the whole scale, range or compass of a thing' (O.E.D.). Perhaps I may be allowed to press the musical analogy further.

God as Creator we now see as, perhaps, somewhat like a composer who, beginning with an arrangement of notes in an apparently simple tune, elaborates and expands it into a fugue by a variety of devices. Thus does a J.S. Bach create a complex and interlocking harmonious fusion of his seminal material both through time and at any particular instant. In this kind of way might the Creator be imagined to unfold the potentialities of the universe which he himself has given it. He appears to do so by a process in which the creative possibilities, inherent (by his own intention) within the fundamental entities of that universe and their interrelations, become actualized within a temporal development shaped and determined by those selfsame inherent potentialities.

The image of creation as an act of composing and of the created order as a musical composition is surprisingly rich and fecund and, since propounding it in my 1978 Bampton Lectures, I have come across a number of other authors resorting to the image of music, as flexible form moving within time, to express what they wish to say about both the created order and the act of creation – authors as diverse as Popper, Capek, and Eigen.[62] The music of creation has also been a constant theme of the religions of India, in particular. It was, indeed, a correct and shrewd instinct on the part of Glansdorff and Prigogine to depict, on the dust-cover of their book expounding the ideas of the Brussels school, a typical South Indian bronze of Shiva, the Creator-Destroyer, as Lord of the Dance of Creation. Within a fiery circle representing the action of material energy and matter in nature, Shiva dances the dance of wisdom and enlightenment to maintain the life of the cosmos and to give release to those who seek him. Coomaraswamy rightly calls attention to (I quote) 'the grandeur of this conception itself as a synthesis of science, religion and art ... In the night of Brahma, Nature is inert, and cannot dance till Shiva wills it: He rises from His rapture, and dancing sends through inert matter pulsing waves of awakening sound, and lo! matter also dances appearing as a glory round about Him. Dancing, He sustains its manifold phenomena.'[63] Dancing involves play and joy, and this conception of the world process as the Lord Shiva's play is also prominent in the Hindu scriptures – 'The perpetual dance is His play.'

Both images, of the writing of a fugue and of the execution of a dance, serve to express the idea of God enjoying, of playing in, creation. This is not an idea new to Christian thought. The Greek Fathers, so Harvey Cox argues, contended that the creation of the world was a form of play. 'God did it they insisted out of freedom, not because he had to, spontaneously and not in obedience to some inexorable law of necessity.'[64]

But this theoretical acceptance of the role of chance in creation and in the created world also has implications for our attitudes to the role of chance in human life. Rustum Roy (Director of the Materials Research Laboratory at Pennsylvania State University) entitled his first 1979 Hibbert Centenary Lecture, in London, 'Living with the Dice-playing God.' He urged us to accept that the world displays patterned chance. It is not a chaos, for there is only a 'loose coupling,' through statistical laws and patterns, which still allows talk of 'causes.' In human life we must accept, for the stability of our own mental health and of our faith, that reality has a dimension of chance interwoven with a dimension of causality – and that through such interweaving we came to be here and new forms of existence can arise. This acceptance of chance as part of the mode of God's creativity is more consistent with the fundamental creativity of reality than the belief – stemming from a Newtonian, mechanistic, determinist view of the universe with God as the great Lawgiver – that God intervenes in the natural nexus for the good or ill of individuals and societies.[65] We must learn to accept these conditions of creation and of creativity in the world – 'the changes and chances of this fleeting world.' Such an attitude can rightly be urged not simply as a psychological necessity but also as the outcome of the recognition that we have just been developing, that the creation of life itself, and the creativity of the living, inevitably involves an interplay of chance and law.

Sociobiology

A new relation between biology and the social sciences? Earlier I discussed evolutionary theory as 'methodologically compositionist,' since it is a view of the biological scene that certainly looks 'from the top down,' that is, it deals primarily with whole organisms in their total environments. One of the all-pervasive problems in the history of evolutionary theory is the identification of the unit that is being selected as subject to evolutionary laws. Nearly all of the possible levels of analysis have, at some time or another, been chosen by some biologist as *the* unit of selection – genes, parts of chromosomes, whole chromosomes, genotypes, organisms, Mendelian populations, biological species, and so on.[66]

In the last few years this area of biology has witnessed sharp controversy emanating from the confrontation of the theory, until then widely accepted by biologists, of 'group selection' with the theory of 'individual, or gene, selection.' As we shall see, the latter theory has largely been expounded by its supporters in a reductionist manner not only in relation to animal behaviour but also in relation to human behaviour, ethics, sociology, and anthropology – and even wider aspects of culture. Hence its classification as exemplifying a reductionist methodology. Broadly, sociobiology may be described as the interdisciplinary study of the biological basis of all social behaviour, and aims at exploring the relationships between biological constraints and cultural evolution. But it has a more particular origin in biology itself as we shall see in the following.

The controversy was initially associated with different interpretations of biologically altruistic behaviour which, for these purposes, may be defined as behaviour by an individual organism of a kind that increases the chances of survival of another like individual, with increased risk to its own survival. 'Survival' is taken here in its Darwinian sense, namely, 'survival in order to reproduce,' and it is now well established that quite small increments in the chance of survival, in this sense, lead surprisingly rapidly to the dominance in biological populations of individuals possessing the genetic factors responsible for this increment. In the 'group selection' theory, altruism was explained on the supposition that a group (for instance, a species, or a population within a species) whose individual members were altruistic was less likely to become extinct than one whose members were non-altruistic, that is, selfish.[67]

But there was a paradox here – for altruistic behaviour reduces the chance of an organism surviving (to reproduce) and so, eventually, organisms that behave thus should disappear from the group, or species. In recent years, a very active group of biologists have resuscitated Darwin's own emphasis on 'individual selection' and now represent altruistic behaviour as *genetic* selfishness. In this theory, what we call altruistic behaviour on the part of an individual, apparently on behalf of other organisms in the group, is simply behaviour which enhances the chance of survival (and so of the reappearance in the next generation) of the genes in those other organisms that they also share with the 'altruistic' individual. So those on behalf of whom the altruistic sacrifice is made must be genetically kin to the altruistic individual. The 'altruism' of, for example, a bird emitting a warning cry to the rest of its kin-group of the approach of a predator, thereby attracting the attack to itself, is simply, on this view, a mechanism for enhancing the chances of survival of genes that are like its own but are carried by those other, related, individuals. We recall that J.B.S. Haldane once affirmed he would lay down his life for two brothers or eight cousins![68] No special 'motivation,' or 'purpose,' or any special awareness of the group, needs to be attributed to the organism – the selection processes and their statistical features ensure this result (the increased chance of reproduction of the genes that the 'altruistic' individual shares with the rest of the group) – and to introduce teleological or group language is simply a post ipso facto gloss on what is actually going on. These ideas have been powerfully argued in E.O. Wilson's monumental work *Sociobiology* and, more popularly, expounded in Richard Dawkins's *The Selfish Gene*.[69]

The argument for individual, or gene, selection as the appropriate interpretative category of behaviour rests on the assumption that one can properly speak of a gene for a particular kind of behaviour, even if we have no knowledge of the actual causal chains linking genes and behaviour.[70] Thus a 'gene for altruistic behaviour' would be one that transmits information that affects the development of the organism's nervous system so as to make it more likely to behave altruistically – and so might have its effect at a number of levels.[71] This way of regarding the role of the 'selfish' gene has been applied to interpreting a wide range of behaviour, other than altruistic – for instance, aggression, the 'battle of the sexes,' parental policies, feeding habits, the relation between old and young. The application involves employing the theory of games to work out what is the most evolutionarily stable strategy, that is, the behavioural policy which, if adopted by most members of a population, cannot be bettered, from the viewpoint of gene *and* population survival, by any other strategy.[72]

In recent years, some sociobiologists, after rebutting the attempts of recent decades to reduce biology to the molecular sciences, have taken upon themselves the role of the unjust steward and have appeared to be attempting to reduce sociology, anthropology, and the sciences of human behaviour to biology. Perhaps this assertiveness should not be taken as an attempt at outright reduction of these sciences to biology, for Wilson has, since the publication of *Sociobiology*, argued for the value to any discipline of its *antidiscipline* (referring to the special, creative, adversary relation that exists initially between the studies of adjacent levels of organization) – with biology as the antidiscipline to the social sciences.[73] Moreover, in the same article he explicitly repudiates any reductionist ambitions of biology with respect to the social sciences, which he recognizes as 'potentially far richer in content' than biology. For Wilson is quite aware that the properties of societies are emergent and hence deserving of 'a special language and treatment'[74] but he, nevertheless, wishes to give a prime and determinative role to the biological basis of human social behaviour and patterns.[75] Such apparent intellectual imperialism has provoked strong reactions from the native denizens of anthropology and sociology – not to mention political opposition which sees sociobiologists as reincarnated nineteenth-century social Darwinists.

One of the weightiest attacks on sociobiology so far published from within one of the 'threatened' sciences is that of Marshall Sahlins, an anthropologist. To Wilson's question of 'whether the social sciences can be truly biologicized in this fashion [of sociobiology],' Sahlins responds: 'The answer I suggest here is that they cannot, because biology, while it is an absolutely necessary condition for culture, is equally and absolutely insufficient: it is completely unable to specify the cultural properties of human behavior or their variations from one human group to another.'[76] For, he argues, 'the central intellectual problem does come down to the autonomy of culture and of the study of culture. *Sociobiology* [E. Wilson's book] challenges the integrity of culture as a thing-in-itself, as a distinctive and symbolic human creation. In place of a social constitution of meanings, it offers a biological determination of human interactions with a source primarily in the general evolutionary propensity of individual genotypes to maximize their reproductive success.'[77] Scientific sociobiologists who attempt to place social behaviour on sound evolutionary principles (notably the self-maximization of the individual genotype) do so, Sahlins suggests, by assuming that human social behaviour can be explained as the expression of those needs and drives of the human organism that have been imprinted by biological evolution.[78] But this position, he claims, does not correspond to the results of anthropological study.[79] As evidence he cites *inter alia* the absence of any relation between war and individual human aggressiveness. The latter may be mobilized to pursue a war but its existence does not in itself explain the existence of war, in general, and the causes of any particular war – 'Aggression does not regulate social conflict, but social conflict does regulate aggression.'[80]

Many sociobiologists (Wilson, Trivers et al.) argue that kin selection – an essentially cost-benefit analysis of an individual's behaviour towards genetic relatives, the 'selfish gene' model – is the deep structure of human social patterns and behaviour. Sahlins, by ranging over the actual arrangements in a number of carefully studied cultures, claims to demonstrate that 'sociobiological reasoning from evolutionary phylogeny to social morphology is interrupted by culture,'[81] so that any claims for sociobiology to be the key to all the human sciences, and indeed all the humanities, are exaggerated.[82]

This attack by a leading anthropologist, whose research results have been used by sociobiologists themselves to support their case, has had considerable influence and cannot, in my view, easily be set aside. Michael Ruse thinks Sahlins's objections are nothing like as devastating as he (Sahlins) thinks because he has ignored two important ideas in sociobiology.[83] One is 'reciprocal altruism' – 'altruistic' action directed at non-genetic kin, on the basis that this evokes reciprocal action which *does* benefit one's genetic kin: 'If I am ready to do it for you, then you are ready to do it for me.' The other is 'parental manipulation' – a form of 'altruism,' of the genetic cost-benefit variety, in which one individual (one offspring) is being manipulated or forced by a second (the parent) to help a third (a sibling of the other offspring), all to help survival of the parents' genes. These two ideas, Ruse claims, can explain, respectively, two kinds of observations that Sahlins thinks are fatal to the sociobiological approach: namely, adoption of enemy's children and infanticide (an admittedly extreme application of parental manipulation in favour of the survival of some siblings at the expense of others). The bother with this counter-attack is that it invokes *both* 'altruism' *and* 'reciprocal altruism' (in the sociobiological senses) – very much a 'fail-safe' intellectual ploy. For a combination of these would allow one to explain *any* behaviour whatever: if it *does* directly benefit one's own genes (now located in some kin), by the 'altruism' interpretation; or if it *does not* benefit them directly, by the 'reciprocal altruism' interpretation. What is needed is actual observational *evidence* that each is the proper explanation in its respective case. Can this combination, which appears to be what Ruse is invoking, now constitute a *scientific* explanation, for *no* evidence can ever falsify these two hypotheses if they are used in conjunction? So the argument continues and it is clear that the plausibility of such sociobiological explanations depends on much more research, as judiciously recognized, for example, by J.H. Crook who clearly believes that 'in principle' such genetic 'biostrategies' are 'at work beneath the surface layer of culture that determines the relationships themselves.'[84]

The controversy has continued sharp and furious not only between sociobiologists and philosophers,[85] and between sociobiologists and anthropologists,[86] but also among biologists themselves[87] – not to mention the political antagonism it has aroused from American groups such as 'Science for the People,'[88] an antagonism almost totally ill founded, in my view, gratuitously attributing, as it does, political motivations to the scientists involved.[89] Crook's summary of the present situation is one of the fairest:

> The tentative explanation[90] of human conduct that stems from the sociobiological paradigm relates man to behavioural and social evolution in the animal kingdom generally and thus for the first time anchors the study of society in evolutionary biology through a fundamental theory. None the less the enormous variety of cultural processes cannot be interpreted solely by sociobiological explanation. Cultures express the attempts of individuals to find meaning in their lives and to produce collectively systems of meaning that make life comprehensible and legitimize action. The capacity to construct interpretative systems rests in the advanced cognitive capacities of man which have evolved in relation to a need to represent social relations in language. The study of what people say in accounting for their actions (emic theory)[91] gives an understanding of the processes of culture while sociobiological theorizing gives an insight into the ultimate meaning of culture itself (etic theory)[92] ... Cultural evolution comprises the historical process which provides the sociological environment within which the basic biological strategies of the species find varied expression.[93]

So we must welcome the real insights biology can provide into the *constraints* which man's biological nature place upon him. That this is the way biology can help in the understanding of man in society has become apparent from the extensive discussions, which began in the nineteenth century, about whether our knowledge of the evolutionary process could generate ethical norms. The conclusion of most philosophers of ethics is that it cannot do so without being guilty of the 'naturalistic fallacy' of deducing what *ought* to be the case from what *is*, though this conclusion needs considerable qualification.[94]

The implications of sociobiology for theology. Sociobiology is concerned with interpreting human social behaviour, with its rich cultural expression and variety, in the light of animal, bird, and insect social behaviour, with their more fixed behaviour patterns (often entirely so in the case of insects) that are described in terms of genetic cost-benefit exchanges. By virtue of thus straddling the world of human culture and that of the behaviour of the non-human biological world, it inevitably touches, indeed sometimes forcibly strikes, upon many issues concerning the fundamental nature of man. The debate about sociobiology is not entirely a replay of the old nature–nurture dichotomy controversy about the factors in human behaviour, because there has been an enormous increase in knowledge of the complexity of the strategy of gene perpetuation and of the many-levelled character of any adequate interpretation of human behaviour (symbolic, psychological, hormonal, neurological, nutritional – not to mention the spiritual, ethical, and intellectual). So many of the issues that the proponents of sociobiology touch upon are those that have again and again been raised by science and philosophy for theology.

The emphatically evolutionary outlook of sociobiology does not, in itself, have any new implications for theology that have not been raised in relation to the general idea of biological (and indeed, cosmic) evolution and have been discussed above (on pages 34–49): namely, questions concerning continuity, chance, emergence, and interconnectedness, with their resulting renewed stress on the immanence of God in the 'natural' processes of creation. However, it is true that the wide-ranging scope of sociobiology and the energy and zest with which its expositors apply and extend it, undoubtedly make even more urgent the need for Christian (and indeed all) theology to become much clearer and explicit about its relation to such views, that is, to the world-view of scientific 'evolutionary naturalism.' This latter approach is the dominant viewpoint of the scientific community today and has been described by Karl Peters in the following terms:

> Evolutionary naturalism may be described as follows: First, the realm of nature is all there is; there is no supernatural in the sense of a realm of knowable reality totally other than that which is open to some possible interpretation of everyday experience by some possible scientific theories. Second, nature is dynamic; it evolves. Change is not merely an appearance or an indication of a second-class reality but is essential to the way things are. Third, at least at the level of life, the evolution of nature is best understood by updated Darwinian mechanisms: a continuing inheritance by the replication of major bodies of information; continual, essentially random, small variations of these information systems; and environmental selection pressures favoring the reproduction of some variations over others and thus modifying in small steps the information heritage.[95]

However incomplete we may regard this view in itself (and Peters's definition of 'no supernatural' needs much qualification) it is one that is extremely well supported on scientific grounds and one whose religious implications require exploration. It will not go away, however uncongenial to traditional theology, and it is increasingly the most widely generally accepted account of at least how we arrived here, if not why. For myself, it is in its bare outline[96] the best account we have of the natural world of which we now know we are part – and sociobiology, stripped of its reductionist overtones, is certainly a new and positive contribution to that evolutionary naturalism. As Peters points out, such evolutionary naturalism is not by itself definitive of any particular theistic, or atheistic, position and is, as a matter of observation, shared by liberal theists, religious humanists, and agnostic and atheistic humanists – if not always by orthodox Christian theologians. But for anyone who believes that the natural world is the sphere of action of God the Creator, it makes new demands upon theological conceptualization, and I have earlier on in this lecture made some attempts to bring to birth some new images.

Sociobiologists are not a uniform group with respect to their philosophical positions, but I think it is fair to say that, by successfully delineating the genetic strategies underlying behaviour patterns and roles in many insect, bird, and animal societies, they have often been confidently and explicitly deterministic, reductionist, and functionalistic in their interpretation of human behaviour – or, perhaps it is more accurate to say, they have shown a general tendency to favour interpretations of human behaviour that have been easily seized upon by those who are determinists, reductionists, and functionalists. Some sociobiologists (for instance, E.O. Wilson and R. Dawkins [in some passages in *The Selfish Gene*]) have gone out of their way to disavow such extreme positions, which other of their writings may have seemed to imply.[97] But the net effect has been a renewed stress on reducing accounts of biological behaviour to a deterministic level that interprets them functionally in terms of their contribution to the survival of genes; behaviour is regarded as a strategy (however indirect) for gene survival. There can be no doubt of the success of many such interpretations in the non-human biological field, but it is over their application to human behaviour that particular controversy arises, as we have seen in the example of kin selection. So the theological response to these ideas, in their general import, is that which must be made to any purely deterministic and reductionistic accounts of human behaviour – along the lines adumbrated on pages 31–3 and to be taken up again in our concluding section. But in making any such appropriate responses, theologians would do well to recognize, more explicitly than they have done in the past, the complexity of human nature and the fact that its basic foundational level is biological and genetic, however overlaid by nurture and culture. And they must couple this recognition with an acknowledgment that it is this kind of genetically based creature that God has actually created as a human being through the evolutionary process. God has made humankind thus with its genetically constrained behaviour – but, through the freedom God has allowed to evolve in such a creature, he has also opened up new possibilities of self-fulfilment, creativity, and openness to the future that require a language other than that of genetics for elaboration and expression.

The scientifically reductionist account has a limited range and needs to be incorporated into a larger theistic framework that has been constructed in response to questions of the kind 'Why is there anything at all?' and 'What kind of universe must it be if insentient matter can evolve naturally into self-conscious, thinking persons?' and 'What is the meaning of personal life in such a cosmos?' Scientists per se are unlikely to seek such incorporation but at least they may be prepared to recognize that the scientific method is not of the kind that can be directed to answering such questions. Meanwhile theologians have to take more seriously the mode of God's actual creation of man through evolution and also our new understanding of the human creature thus formed – even though, in the past, words such as 'determinism,' 'reductionism,' and 'functionalism' have been red rags to the theological bull! For the genetic constraints upon our nature and action are, from a theistic viewpoint, what God has determined shall provide the matrix within which freedom can operate. But is not this issue nothing other, in a new form, than the old theological chestnut of predestination and free will? Where the Christian theist differs from the sociobiologist, as such, is in his affirmation of *God* as 'primary cause' or ground of being of the whole evolutionary process and, indeed, of God as the agent in, with, and under this process of creation through time. It is the new apprehension and explication of God's presence and agency in the processes that biology in general, and sociobiology in particular, have unveiled that constitute the challenge to theology. Of course, many sociobiologists will be opposed to setting their science in such a wider, theistic framework; for E.O. Wilson, for example, 'no species, our own included, possesses a purpose beyond the imperatives created by its genetic history ... If the brain evolved by natural selection, even the capacities to select particular esthetic judgments and religious beliefs must have arisen by the same mechanistic processes,' and 'scientists cannot in all honesty serve as priests.'[98]

Here conflict between theology and a particular philosophical interpretation of biology is inevitable but the theologian should not enter the lists with destructive ambitions. Indeed some theologians have even argued that theology must come to terms with the domination of the biological process by the prime requirement for *survival*, whether it be of genes, individuals, groups, or species. Philip Hefner, for example, argues that, in the light of biological evolution in general and the sociobiological critique in particular, the whole discussion of the *is/ought* dichotomy – which, as the 'naturalistic fallacy,' has for too long (he thinks) prevented us from seeing how the biological process generates human values – has moved into the arena of survival and non-survival.[99] He goes on to use the categories of A.J. Dyck to elaborate the 'ought' as 'moral requiredness' which is described as a 'gap-induced requiredness': moral requiredness is a gap that we feel compels us (moral obligation) to act, so as to fill it, in order to improve some situation.[100] So Hefner then argues that 'the most urgent gap experienced by humans [in relation to its value-requirements] – and therefore the most pressing gap-induced requiredness – is the gap created by the possibility of not surviving. Theology, therefore, has no alternative but to speak its truth about what is and ought to be in terms relevant to survival – the survival of the species, of the world, *of values, of human worth, of all the conditions upon which the human spirit is dependent*'[101] (my italics).

But the question is: Can the values, and so on, whose survival is to be spoken of in the italicized end of this quotation, be regarded as desirable simply from contemplation of the sociobiological facts (if they *are* facts)? Mary Hesse comments on this prescription of theology's task by Hefner as follows: 'But whatever facts may be discovered about the conditions of survival by sociobiology, the conclusion that the survival of the human species is the most urgent *value* may itself be regarded as *morally* repugnant. This is surely a sufficient rebuttal of the claim that the facts alone permit the "ought" to be derived from the "is" ... God in his wisdom may have ordained values which are consistent with earthly extinction; to suppose otherwise is to embrace some form of materialism.'[102] Whether or not this is a 'sufficient rebuttal' will certainly be argued, but I quote this interchange as an example of a new kind of question (re 'survival' and what it means) that is raised for theology by sociobiology.[103]

There is an application of sociobiology that is relevant to theology and that has been taken up by a number of evolutionary naturalists sympathetic to religion in general, if not especially to Christian theology as such. This is the view that the religions have had a function in enabling human societies (and genes?) to survive and, to that extent, can be justified as useful, functional mythologies – even if they are now ripe, according to E.O. Wilson, for replacement by 'the evolutionary epic' as 'probably the best myth we will ever have.'[104] D.T. Campbell and R.W. Burhoe both argue for a positively selective role for religion in the survival of cultures (their unit of survival, and so of selection); and Burhoe especially, unlike Wilson, argues for its continuing role in the development and survival of human culture, providing it can incorporate the scientific world-view.[105] No doubt Christian theologians will be grateful for this attribution of a survival function to religion in human culture, but the attribution again raises the questions of 'Survival for What?' 'Is survival a *value*?' and 'What *kind* of survival?' – and theologians would (or *should*, in my view) be first asking questions about the *truth* of their formulations, regardless of their contribution to the survival of human culture(s). And could one not argue that it is the ultimate commitment, without regard to survival calculations, to the *truth* that is in God and his Christ that characterizes faith – think of Job's 'Though he slay me, yet will I wait for him.'[106] Is not that also the core of a religion that has a cross as its central symbol and historical focus?[107]

Whatever the outcome is of this particular argument concerning the relation to theology of the central role of 'survival' in biology, what we have witnessed in the last few years is a spate of publications from those who wish to emphasize, in a positively holistic fashion, the physical and biological rooting of the mental and spiritual lives of human beings.[108] In varying degrees, these works see human mental and spiritual life as continuous with, and as a development and elaboration of, the physical and biological (especially genetic) substratum through which evolution has operated. We have also witnessed the recognition by at least one eminent biologist, Sir Alister Hardy, of the religious experience of human beings as one of their natural characteristics and as amenable to scientific investigation, at least in the style of 'natural history.'[109] Our mental and spiritual life, it seems, must fulfil at least basic evolutionary requirements long established, whatever terms it may have to use to interpret itself to itself at its own level. So the pressure from the ideas of sociobiology, in particular, and those of biology and cosmic evolution, in general, is towards a franker recognition of our natural relatedness to the physical and biological worlds and an acknowledgment that our mental and spiritual aspirations *are* rooted therein. But what those aspirations should be directed to is not thereby prescribed and so it is that theology has, in my view, a new and exciting role to play if it will only recognize its new brief.

A NEW STIMULUS FOR THEOLOGY

In the history of the people of Israel, the good Lord was always raising up apparent scourges, such as Cyrus, that were blessings in disguise to lead his people through the trauma that alone would enable them to apprehend new truths.[110] Thus it may be with the new biology – and indeed with the larger perspective of all the sciences – and our comprehension and apprehension of the God who is always greater than that which we can at any time conceive. I have already discussed how the new biology, in the evolutionary setting which is afforded it by cosmology and the earth sciences, has implications for any theology of nature, man, and God. The new biology raises questions for us in a new way, with a new shape and in a new context, and if theological affirmations are not to be meaningless – the mere inner musings of a religious ghetto – they simply have to respond to the questions so raised. They must respond in the terms in which the new knowledge raises them, and not in the language of the intellectual framework that originally provided the setting for classical

Christian dogmatic formulations, whether that of the late Roman Empire, or sixteenth-century Germany, or nineteenth-century England.

To have the questions raised in this way is a new stimulus, in my view, for theology. For when the biological, and other, sciences prompt these questions they do so in a way that reveals they cannot answer them with the resources of the scientific method and intellectual procedures. So that, at last, theology has a chance to respond to questions that might actually be being asked within the context of our present scientific culture. In responding, I think theology will find there is scarcely any one of the main 'heads' of Christian doctrine, as they used to be called, that is not affected by the new perspective. Perhaps one day a new systematic theology might emerge prompted by this stimulus and so fulfil, or at least continue, the quest of William Temple which I described at the start.

Let me now, with ludicrous brevity, indicate the general style of what I think should be our response to these new stimuli, under the 'heads' of William Temple's own magisterial treatment.

Nature we now see as multi-levelled and hierarchical with new emergents developing in time among which, significantly, are consciousness and self-reflective self-consciousness. The natural physical, biological, human, and social worlds are the realm of God's immanent activity, indeed the manifestation of his creative presence – and sacraments are particular focused instances of what is happening all the time. For one should speak of the world, of nature, of all-that-is, as being 'in God,' rather than of God as 'in the world.' Nothing happens apart from God, but God is more and other than what happens. Awareness of the continuity of the physical, non-biological, and biological with the distinctive life of man and the realization that our mental and spiritual life is an emergent from this natural matrix provide a vision in which the whole world takes on a new value and significance – not simply because it resulted in us, but because it *is* an expression of God in action in the way appropriate to the levels in question, and we are that part of the cosmos consciously capable of being aware of and of responding to that immanent Presence. Man's most intellectual and spiritual activities are part of the created order and an aspect of them. 'That there should "emerge" in the cosmic process a capacity to apprehend, even in a measure to comprehend, that process is the most remarkable characteristic of the process itself ... That the world should give rise to minds that know the world involves a good deal concerning the nature of the world.'[111] This assessment and way of raising the question of God finds, in my view, increasing and ample justification in today's biological perspective.

Man. The scientific fact that matter, after a succession of levels of self-transcendence, can in man become self-conscious and personal, self-transcendent, and corporately self-reflective is a fundamental feature of the cosmos and must be regarded as a clue to its meaning and intelligibility. So the capacity for consciousness, and then self-consciousness gradually and increasingly manifests itself immanently within the natural order and reveals itself for what it is; and thereby what it also becomes capable of revealing is the immanence of the transcendent Creator – 'In the beginning was the Word ... and the Word was God ... all things were made by him ... in him was life and the life was the light of men.'[112] But that light shone in darkness and was not comprehended. Man's potentialities – qualitatively so far removed from even those of the highest mammal in respect to his possible creation of value, beauty, truth, and goodness – are not fulfilled and he is conscious of, indeed self-reflective on, his need to come to terms with death, finitude, and suffering. In man, it almost appears as if the evolutionary process has faltered, for man's tragedy is both to be aware of his potentialities and to be unable to attain them. He needs redemption, not in the sense of any time-reversal so that a pristine state of original innocence can be restored, but so that he can – with his peculiar combination of self-consciousness, freedom to choose the good, and biological limitations – yet fulfil his potentialities and attain harmony with his Creator and so, *pari passu*, with the created natural world. In my view of the implication of the evolutionary perspective, it is with reference to such a placing of man that the 'redemption' made possible in Christ will henceforth have to be proclaimed. The role in man's redemption of explicit sacraments in a sacramental universe in which the spiritual is known *only* 'in, with, and under' the physical stuff of existence will clearly need further exploration.

God, who is the Transcendent One immanent in the creative process, can no longer be regarded as the sublimely indifferent Absolute remote from all the pangs of the creative process. For in that process new life, new forms, only emerge through death of the old, and suffering and death are frequently the only gateway to new life and new creation. So we have to conceive of God suffering in and with and, indeed, through creation. When ultimately the transcendence-in-immanence and immanence-in-transcendence are finally fused in a manifestation that is both human and personal, that manifestation in the Word-made-flesh himself goes through the door of suffering and death to fullness of life and the consummation of humanity within the presence of God – so the final agony and apogee of the evolutionary process is a man on a cross exalted into the life of God.

Thus it is that we come to where Temple began and ended, to that Unity in Trinity after which this college is named, to that mystery of the union of God the Father, Son, and Holy Spirit – of Transcendence and Incarnation and Immanence – and to the experiment of life in which that unity may be experienced even if beyond expression and beyond all words. No science, not even the science of natural life, will take us that far, but at least scientific experiments *on* life can help to inform us accurately of the parameters within which we all have to make the 'experiment *of* life.'

NOTES

1 William Temple, *Nature, Man and God*, 1932–33 and 1933–34 Gifford Lectures (London: Macmillan 1935), lecture 3, 57

2 Michael Ramsey, *From Gore to Temple* (London: Longman 1960), 161

3 Quoted by Ramsey, *From Gore to Temple*, 161.

4 *From Gore to Temple*, 160

5 Temple, *Nature, Man and God*, 474–5

6 In *Lux Mundi*, ed. C. Gore (London: Murray 1889)

7 *Nature, Man and God*, 52–3

8 'It may safely be said that one ground for the hope of Christianity that it may make good its claim to be the true faith lies in the fact that it is the most avowedly materialist of all the great religions. It affords an expectation that it may be able to control the material, precisely because it does not ignore it or deny it, but roundly asserts alike the reality of matter and its subordination. Its own most central saying is: "The Word was made flesh," where the last term was, no doubt, chosen because of its specially materialistic associations. By the very nature of its central doctrine Christianity is committed to a belief in the ultimate significance of the historical process, and in the reality of matter and its place in the divine scheme' (*Nature, Man and God*, 478).

9 'It is clear that ... we are trying to frame a conception which is not identical with any of the commonly offered suggestions concerning the relation of the eternal and the historical, and are now extending its application so as to include the relation of the spiritual and material. It is not simply the relation of ground and consequent, nor of cause and effect, nor of thought and expression, nor of purpose and instrument, nor of end and means; but it is all of these at once. We need for it another name; and there is in some religious traditions an element which is, in the belief of adherents of those religions, so close akin to what we want that we may most suitably call this conception of the relation of the eternal to history, of spirit to matter, the sacramental conception' (*Nature, Man and God*, 481–2). See also Dr D.R.G. Owen's article in this volume.

10 For further amplification see A.R. Peacocke, 'Matter in the Theological and Scientific Perspectives – a Sacramental View,' in I.T. Ramsey ed, *Thinking about the Eucharist* (London: S.C.M. Press 1972), 14–37.

11 For a fuller exposition see the author's review on reductionism in *Zygon* 11 (1976): 306–34 and a shorter statement in chapter 3, *Creation and the World of Science* (Oxford: Clarendon Press 1979).

12 This *methodological* distinction – a useful one – was made, inter alia, by T. Dobzhansky in relation to biology in S. Morgenbesser, P. Suppes, and M. White eds, *Philosophy, Science and Method* (New York: Macmillan 1969), 165–78.

13 F.H.C. Crick, *Of Molecules and Man* (Seattle: University of Washington Press 1966), 10

14 Following M. Beckner, 'Reduction, Hierarchies and Organism,' in A.J. Ayala and T. Dobzhansky eds, *Studies in the Philosophy of Biology* (London: Macmillan 1974), 63–76.

15 In the precise ways that philosophies of science, such as that of E. Nagel (*The Structure of Science* [New York: Harcourt, Brace and Co. 1961], chap. 11), describe.

16 W. Wimsatt, 'Reductionism, Levels of Organisation and the Mind–Body Problem,' in G.G. Globus, G. Maxwell, and I. Savodnik eds, *Consciousness and the Brain: A Scientific and Philosophical Inquiry* (New York: Plenum 1976), 199–267; W.C. Wimsatt, 'Robustness, Reliability and Multiple-determinism in Science,' in M. Brewer and B. Collins eds, *Scientific Inquiry and the Social Sciences: A Volume in Honour of Donald T. Campbell* (San Francisco: Jossey Bass 1981)
17 See note 12.
18 F. Jacob, *The Logic of Living Systems* (London: Allen Lane 1974), 13
19 *Nature* 289 (1981): 735; 290 (1981): 75–6, 82
20 'Cladism' is based on the assumption that only shared, derived homologies indicate branching relationships within any group of organisms and that classification of any such group of organisms should be directly related to the branching diagram ('cladogram') indicating phylogenetic affinities. It has been expounded recently by N. Eldredge and J. Cracraft in their *Phylogenetic Patterns and the Evolutionary Process* (New York: Columbia University Press 1980). Its tenets, according to A.J. Cain, who sharply criticized it in an address to the 1981 British Association Meeting at York (reported in *Nature* 293 [1981], 15–16), are: 'The amount of difference between forms is directly proportional to the age of their common ancestor (the further apart, the earlier the ancestor) as if rates of evolution were constant. All evolution is dichotomous, so that if for any one group one can identify its sister group, then their common ancestor can be constructed according to a set rule. And, most remarkable of all, convergence has not occurred unless it can be shown to have occurred; this is said to be a heuristic principle, without which the basis would be removed from phylogenetic systematics' (p. 16).
21 C. Patterson, letter to *Nature*, 288 (1980): 430 – replying to a letter of L.B. Halstead, ibid., 208
22 See the full report by R. Lewin in *Science* 210 (1980): 883–7.
23 S.J. Gould and N. Eldredge, in J.W. Schopf ed, *Models in Palaeobiology* (San Francisco: Freeman 1972); idem., *Palaeobiology* 3 (1977): 115
24 *The Living Stream* (London: Collins 1965), 161ff, 189ff
25 L.B. Halstead, *Nature* 288 (1980): 208
26 See J. Maynard-Smith, *The Theory of Evolution*, 2nd ed. (London: Pelican Books 1966) for one of the best accounts of these subtleties. An excellent contemporary (1981) assessment of the question 'Is a new evolutionary synthesis necessary?' by G.L. Stebbins and F.J. Ayala, has recently appeared in *Science* 213 (1981): 967–71. They conclude that the 'modern synthesis' does not preclude the kind of emphasis made by the 'punctualists,' and their article is notable for a particularly lucid perception of the relation of different levels of theory (from the *micro*- to the *macro*-evolutionary) which are

assessed in relation to the issues concerning reduction, along lines very similar to those outlined above on pages 31–3 (and stated more fully in the works cited in note 11). See also the judicious assessment, concerning the speed of evolution, of J.S. Jones in *Nature* 293 (1981): 427.

27 R.M. Gorczynski and E.J. Steele, *Nature* 289 (1981): 678. See also R.B. Taylor, *Nature* 286 (1980): 837; an article entitled 'Too Soon for the Rehabilitation of Lamarck,' in *Nature* 289 (1981): 631–2; and an account (R. Lewin, *Science* 213 [1981]: 316–21) of the whole subsequent controversy about the interpretation of these and other experimental observations, and of the personal conflicts involved.

28 See, for instance, R. Duncan and M. Weston-Smith eds, *The Encyclopaedia of Ignorances* (Oxford: Pergamon Press 1977).

29 See note 13.

30 R. Sheldrake, *A New Science of Life* (London: Bland and Briggs 1981)

31 *Nature* 290 (1980): 82

32 *Logic of Living Systems*, 13

33 See M.O. Dayhoff, *Atlas of Protein Sequence and Structure* (Washington, DC: National Biomedical Research Foundation 1972), vol. 5, 8 and D55.

34 Allowance has to be made for variations within a given species.

35 G.G. Simpson, *The Meaning of Evolution* (New Haven: Bantam Books, Yale University Press 1971), 236

36 Simpson, *Evolution*, 258–9

37 More fruitful may be a suggestion of K. Denbigh (in *An Inventive Universe*, [London: Methuen 1975], 98ff) who has pointed out that, although we have concepts of orderliness, or order, and of disorder, we do not have any satisfactory measure of *organization*. He has therefore proposed as a measure of complexity a quantity he names *integrality*, which is the product of the number of connections in a structure and the number of different *kinds* of parts. Integrality is not identical with 'information,' or with entropy; it can increase in a closed system (for example, when an egg develops into, say, a chick) and its total value on the earth has increased since life began.

38 Pierre Teilhard de Chardin, *The Phenomenon of Man* (London: Collins 1959), 300–2

39 It has recently been argued (John Baker, *Expository Times* 92 [1981]: 235–7) that the story of Adam and Eve has a different significance *as myth* than that traditionally attributed to it: 'Before the entry of the serpent there is a harmonious relationship between Adam and Eve and God. In the calm and plenty of the Garden no question of need or doubt arises. Most important of all, Adam and Eve are in a state of absolute acceptance, of innocence, and the *possibility of choice* has never arisen. What the serpent achieves is to pose an alternative to innocence, to introduce the possibility of *choice*. The

theological doctrine of the Fall argues that the serpent's (i.e., Satan's) aim was to alienate man from God. Within the setting of the myth, however, all that it could do was to offer an alternative to innocence, without necessarily causing a rift between man and God. On this basis the story of Adam and Eve takes on a different significance. What happens there is not a "Fall", but an *awakening*, and the so-called "alienation" or separation of man and God is really a form of freedom, necessary to man's full development. Man should have knowledge and choice and power, if he is to be fully man. When looked at without the bias of later theology, the myth is seen to be the story of man's inevitable spiritual development, out of dependence, out of innocence and total security, into the world of reality and moral choice. It describes the loss of one kind of harmony – a childlike identity with God – but what it does not point out is the possibility of another kind of harmony between man and God, based not on an unconscious innocence, an identity of man with God (similar to that of animals with nature), but a harmony arising from choice, i.e., the adult, self-conscious man choosing without compulsion the will of God in perfect freedom, and doing this even in the face of temptation and stress. Thus the story of the Garden portrays, not man's Fall, but man's liberation, his entry into full adulthood, possessed not of unconscious goodness and incorruptible innocence, but of the power of choice – i.e., "the knowledge of good and evil" ... What of the later Christian doctrine of the "Fall"? This was based on man's "disobedience" of God's command, but with a complete lack of understanding on the part of the early theologians of the nature of that disobedience. They did not consider the alternative – the state of innocence, or else they were so enamoured of that alternative that they failed to understand its true nature. The myth really concerns man's choice of free-will, as opposed to the blissful innocence of the un-free robot ... In those conditions, the choice of freedom cannot be called "sin", nor can the resultant state be called a "Fall". The true story of man's Fall could, for a theologian determined to find it in the Bible record, only begin with the sin of Cain, who exercised his power to do an evil deed' (pp. 236–7).

40 For a more detailed statement of this view, see the author's *Science and the Christian Experiment* (London: Oxford University Press 1971), 148ff, and *Creation and the World of Science* (Oxford: Clarendon Press 1979), 190ff.

41 F. Capra, *The Tao of Physics* (London: Collins, Fontana 1976)

42 See chapter 7 of the author's *Creation and the World of Science*.

43 The whole area of brain research and the neuro-sciences is developing so rapidly and involves such a vast new philosophical literature on the body-mind problem that it has been impossible to include any assessment of it in this section.

44 G.S. Stent, *Science* 160 (1968): 390–5
45 Ibid., 391
46 N. Bohr, address on 'Light and Life' to the International Congress of Light Therapy in 1932 – *Nature* 131 (1933): 421
47 W. Astbury, *Harvey Lectures*, 1950–51 (Springfield, Ill.: Thomas 1952), 3
48 Stent, *Science* 160: 391
49 J.D. Watson, *The Double Helix* (London: Weidenfeld and Nicolson 1968); R. Olby, *The Path to the Double Helix* (London: Macmillan 1974); H.F. Judson, *The Eighth Day of Creation* (London: Jonathan Cape 1979)
50 Stent concludes: 'I think it fair to say, by way of appreciation of the dogmatic phase, that there have been two great theories in the history of biology that went more than a single step beyond the immediate interpretation of experimental results; these were organic evolution and the central dogma' (*Science* 1960: 394).
51 S. Brenner, *Nature* 248 (1974): 787
52 G.S. Stent, *Nature* 248 (1974): 779–81
53 K.F. Schaffner, *Science* 157 (1967): 644–7
54 Re Beckner, see note 14 above. For a survey, see A.R. Peacocke, note 11.
55 I. Prigogine in *From Being to Becoming* (San Francisco: W.H. Freeman and Co. 1980): 'There is a microscopic formulation that extends beyond the conventional formulations of classical and quantum mechanics and *explicitly* displays the role of irreversible processes. This formulation leads to a unified picture that enables us to relate many aspects of our observations of physical systems to biological ones. The intention is not to 'reduce' physics and biology to a single scheme, but to clearly define the various levels of description and to present conditions that permit us to pass from one level to another' (pp. xiii, xiv). 'We start with the observer, a living organism who makes the distinction between the future and the past, and we end with dissipative structures, which contain, as we have seen, a "historical dimension". Therefore, we can now recognize ourselves as a kind of evolved form of dissipative structure and justify in an "objective" way the distinction between the future and the past that was introduced at the start. Again *there is in this view no level of description that we can consider to be the fundamental one. The description of coherent structures is not less "fundamental" than is the behavior of the simple dynamical systems'* (p. 213, my italics). And D.C. Mikluckecky, W.A. Wiegand, and J.S. Shiner, 'A Simple Network Thermodynamic Method for Modeling Series-Parallel Coupled Flows. I. The Linear Case,' *J. Theor. Biol.* 69 (1977): 471–510: 'The main question of interest ... is whether the network [thermodynamic] approach will be of any help in resolving this philosophical dilemma [of reductionism *versus* 'wholism,' namely that biology deals with a level of organization not found among the

objects of study in physics and chemistry]. To the extent that it provides a method for dealing with more complicated organizational patterns, it must. On the other hand, reducing a living system to a network is not far from reducing it to a collection of molecules. *The networks, as models, are more models of our theories and hypotheses about how the living system works than of the living system itself*. By creating the appropriate network, various notions we have about the workings of an organism can be quantitatively tested' (p. 510, italics mine).

56 This is clearly also the view of F. Jacob, one of the architects of molecular biology in its crucial developments in the 1960s: 'Biology has demonstrated that there is no metaphysical entity hidden behind the word "life". The power of assembling, of producing increasingly complex structures, even of reproducing, belongs to the elements that constitute matter. From particles to man, there is a whole series of integration, of levels, of discontinuities. But there is no breach either in the composition of the objects or in the reactions that take place in them; no change in "essence". So much so, that investigation of molecules and cellular organelles has now become the concern of physicists ... This does not at all mean that biology has become an annex of physics, that it represents, as it were, a junior branch concerned with complex systems. At each level of organization, novelties appear in both properties and logic. To reproduce is not within the power of any single molecule by itself. This faculty appears only with the simplest integron deserving to be called a living organism, that is, the cell. But thereafter the rules of the game change. At the higher-level integron, the cell population, natural selection imposes new constraints and offers new possibilities. In this way, and without ceasing to obey the principles that govern inanimate systems, living systems become subject to phenomena that have no meaning at the lower level. *Biology can neither be reduced to physics, nor do without it*. Every object that biology studies is a system of systems. Being part of a higher-order system itself, it sometimes obeys rules that cannot be deduced simply by analysing it. This means that each level of organization must be considered with reference to the adjacent levels. It is impossible to understand how a television set works without first knowing how transistors work and then something about the relations between transmitters and receivers. At every level of integration, some new characteristics come to light. As physicists already observed at the beginning of the twentieth century, discontinuity not only requires different means of observation; it also modifies the nature of phenomena and even their underlying laws. Very often, concepts and techniques that apply at one level do not function either above or below it. The various levels of biological organization are united by the logic proper to reproduction. They are distinguished by the means of communication, the

regulatory circuits and the internal logic proper to each system' (*Logic of Living Systems*, 306–7). (An 'integron' is each of the units, in a hierarchy of discontinuous units, formed by the integration of sub-units of the level below. An integron is formed by assembling integrons of the level below it; it takes part in the construction of the integron of the level above. [Ibid., 302])

57 J. Monod, *Chance and Necessity* (London: Collins 1972)

58 Ibid., 110

59 I. Prigogine and G. Nicolis, *Quart. Rev. Biophysics* 4 (1971): 132

60 M. Eigen and R. Winkler, *Laws of the Game* (New York: Knopf 1981)

61 M. Eigen, *Naturwissenschaften* 58 (1971): 519

62 K. Popper, *The Unended Quest* (London: Collins [Fontana] 1976); Popper also quotes Kepler to this effect. M. Capek, *The Philosophical Impact of Contemporary Physics*, 1961. For Eigen, see note 60.

63 A.K. Coomaraswamy, *The Dance of Shiva* (London: Peter Owen 1958), 70

64 Harvey Cox, *The Feast of Fools* (Cambridge, Mass.: Harvard University Press 1969), 151

65 Cf. 'Two men will be in the field; one is taken and one is left' (Matt. 24:40); 'Does not God send his rain on the just and the unjust?' (Matt: 5:45). And do *you* think the eighteen who were killed by the fall of the Tower of Siloam were 'offenders above the men that dwell in Jerusalem?' (Luke 13:4).

66 See May B. Williams, 'The Logical Status of Natural Selection and Other Evolutionary Controversies: Resolution by an Axiomatization,' in *The Methodological Unity of Science* (Dordrecht: Reidel Publ. Co. 1973).

67 V.C. Wynne-Edwards, *Animal Dispersion in Relation to Social Behaviour* (Edinburgh: Oliver and Boyd 1962); R. Ardrey, *The Social Contract* (London: Collins 1970)

68 J.B.S. Haldane, 'Population genetics,' *New Biology* 18 (1955): 34–51

69 E.O. Wilson, *Sociobiology – The New Synthesis* (Cambridge, Mass.: Belknap Press, Harvard University Press 1975). R. Dawkins, *The Selfish Gene* (Oxford: Oxford University Press 1976)

70 Though, on any reckoning, it would have to be a *system* of concomitantly acting, linked genes – that is, inheritance of behavioural characteristics is likely to be polygenic. So in the text, as in the Dawkins quotation, take the singular 'gene' to refer to such a system.

71 'For purposes of argument it will be necessary to speculate about genes 'for' doing all sorts of improbable things ... We are saying nothing about the question of whether learning, experience, or environmental influences enter into the development of the behaviour. All you have to concede is that it is possible for a single gene, other things being equal and lots of other essential genes and environmental factors being present, to make a body more likely to save somebody from drowning than its allele would' (Dawkins, *Selfish Gene*, 66).

72 Dawkins summarizes this way of looking at biological evolution and the behaviour of living organisms thus: 'Replicators began not merely to exist, but to construct for themselves containers, vehicles for continued existence. The replicators which survived were the ones which built *survival machines* for themselves to live in. The first survival machines probably consisted of nothing more than a protective coat. But making a living got steadily harder as new arrivals arose with better and more effective survival machines. Survival machines got bigger and more elaborate, and the process was cumulative and progressive. Was there to be any end to the gradual improvement in the techniques and artifices used by the replicators to ensure their own continuance in the world? There would be plenty of time for improvement. What weird engines of self-preservation would the millennia bring forth? Four thousand million years on, what was to be the fate of the ancient replicators? They did not die out, for they are past masters of the survival arts. But do not look for them floating loose in the sea; they gave up that cavalier freedom long ago. Now they swarm in huge colonies, safe inside gigantic lumbering robots, sealed off from the outside world, communicating with it by tortuous indirect routes, manipulating it by remote control. They are in you and in me; they created us, body and mind; and their preservation is the ultimate rationale for our existence. They have come a long way, those replicators. Now they go by the name of genes, and we are their survival machines ... We are survival machines, but "we" does not mean just people, it embraces all animals, plants, bacteria, and viruses' (Dawkins, 21–2)

73 E.O. Wilson, 'Biology and the Social Sciences,' *Daedalus* 106, no. 6 (Autumn 1977): 127–40

74 Wilson, *Sociobiology*, 7

75 'Sociobiology is defined as the systematic study of the biological basis of all social behavior. For the present it focuses on animal societies, their population structure, castes, and communication, together with all the physiology underlying the social adaptations. But the discipline is also concerned with the social behavior of early man and the adaptive features of organization in the more primitive contemporary human societies ... It may not be too much to say that sociology and the other social sciences, as well as the humanities, are the last branches of biology waiting to be included in the Modern Synthesis. One of the functions of sociobiology, then, is to reformulate the foundations of the social sciences in a way that draws these subjects into the Modern Synthesis. Whether the social sciences can be truly biologicized in this fashion remains to be seen' (p. 4).

76 Re Wilson, see previous note; M. Sahlins, *The Use and Abuse of Biology* (Ann Arbor, Mich.: University of Michigan Press 1976), xi

77 Ibid., x. For a thoroughgoing attempt to interpret cultural 'evolution' in terms based on the mathematical theory of biological evolution, see L.L. Cavalli-

Sforza and M.W. Feldman, 'Towards a Theory of Cultural Evolution,' *Interdisciplinary Science Review* 3 (1979): 99–107, and their *Cultural Transmission and Evolution* (Princeton, NJ: Princeton University Press 1981), reviewed by C.R. Cloninger in *Science* 213 (1981): 858–9.

78 '[In sociobiology] The chain of biological causation is accordingly lengthened: from genes through phenotypical dispositions to characteristic social interactions. But the idea of a necessary correspondence between the last two, between human emotions or needs and human social relations, remains indispensable to the scientific analysis ... The interactions of organisms will inscribe these organic tendencies [aggressiveness, altruism, male 'bonding,' sexuality, etc.] in their social relations. Accordingly there is a one-to-one parallel between the character of human biological propensities and the properties of human social systems ... For him [E.O. Wilson], any Durkheimian notion of the independent existence and persistence of the social fact is a lapse into mysticism. Social organization is rather, and nothing more than, the behavioral outcome of the interaction of organisms having biologically fixed inclinations. There is nothing in society that was not first in the organisms' (Sahlins, *Use and Abuse*, 4–5).

79 For 'the problem is that there is no necessary relation between the phenomenal form of a human social institution and the individual motivations that may be realized or satisfied therein. The idea of a fixed correspondence between innate human dispositions and human social forms constitutes a weak link, a rupture in fact, in the chain of sociobiological reasoning' (Ibid., 7).

80 Ibid., 9

81 Ibid., 11

82 Sahlins argues against sociobiology's pre-eminence on the grounds that (1) no system of human kinship relations is organized in accord with the genetic coefficients of relationship as known to sociobiologists; (2) the *culturally constituted* kinship relations, which govern production, property, mutual aid, and marital exchange, have an entirely different calculus from that predicted by genetic kin selection; (3) kinship is a unique characteristic of human societies, distinguishable precisely by its freedom from natural (genetic) relationships; (4) human beings reproduce not as physical or biological beings but as *social* beings, that is, human reproduction is engaged as the means for the persistence of co-operative social orders not vice versa, and so, finally, (5) culture is the indispensable condition of systems of human organization and reproduction. For, he would argue, 'Human society is cultural, unique in virtue of its construction by symbolic means' (p. 61) and 'Culture is biology plus the symbolic faculty' (p. 65); the importance of the symbolic is to generate meaning not merely to convey information, as Wilson seems to want to say.

83 M. Ruse, *Sociobiology: Sense or Nonsense*? (Dordrecht-Boston: Reidel 1979), 122–6

84 J.H. Crook, *The Evolution of Human Consciousness* (Oxford: Clarendon Press 1980), 162ff. In Crook's view: 'Sociobiology explains why human behavior is not arbitrary, why it is structured in a broadly characteristic way wherever people are, but it does not proceed to reduce all descriptions of individual human action to biological causation ... Cultural evolution thus comprises a historical process that provides human beings with the sociological environment within which the basic biological strategies of the species are worked out' (pp. 186–7).

A similar view is expressed also by Pierre L. van den Berghe in *Human Family Systems, An Evolutionary View* (New York: Elsevier 1979) but in terms more critical of the purely cultural interpretations he attributes to Sahlins. 'The question is thus not who [the social and cultural anthropologists or the sociobiologists] is right and who is wrong, but the level of explanation sought. Sociobiology does very poorly in explaining conscious motivations of individuals, and the detailed cultural idiom through which a behavior is expressed, such as the ceremonies of a wedding ritual. On the other hand ... sociobiology goes a long way in accounting for basic structural features of human societies ... Human systems of kinship and marriage conform to only a few basic types. Underlying a great deal of variety in detail, human societies share much of their basic structure of kinship and marriage. The variations, while they attest to our versatility in adapting to a wide range of environmental conditions, are themselves not random but adaptive' (p. 87). 'The modalities of our behavior are indeed importantly shaped by the cultural rules we invent for ourselves; but the rules themselves reflect an underlying biological reality. Human behavior in general, and mating and reproductive behavior in particular, are thus *both* cultural and biological, *both* genetically and environmentally determined, and it makes no sense to dissociate one aspect from the other. What we call kinship and marriage in humans is the cultural expression of our biologically predisposed system of mating and reproduction. We set up societies around mating and reproduction and we cook up rules of conduct about them because it was adaptive for us to do so. The few groups foolish enough to do otherwise failed to survive in competition with those that did. This may not remain true for all time, but it has been true everywhere so far; and "everywhere so far" is all the data we have with which to construct an explanatory social science' (pp. 88–9).

85 See, for instance, Mary Midgley, *Beast and Man: the Roots of Human Nature* (London: Methuen 1980), and her swingeing attack on Dawkins's *The Selfish Gene* in *Philosophy* 54 (1979): 439–58; S. Hampshire's review in *The New York Review of Books*, 12 Oct. 1978, pp. 64–9, of E.O. Wilson's *On Human Nature* (Cambridge, Mass.: Harvard University Press 1978).

86 Sahlins, *Use and Abuse*, and E. Leach's review, entitled 'Biology and Social Science: Wedding or Rape?' (*Nature* 291 [1981]: 267–8), of *Genes, Mind and*

Culture by C.J. Lumsden and E.O. Wilson (Cambridge, Mass.: Harvard University Press 1981) – also critically reviewed from a more genetic viewpoint by C.R. Cloninger and S. Yokoyama in *Science* 213 (1981): 749.

87 See the appropriate sections in A.L. Caplan ed, *The Sociobiology Debate* (New York: Harper and Row 1978); and Ruse, *Sociobiology*.

88 See Caplan ed, *Sociology Debate*, 280–90.

89 See Wilson's rejoinder in Caplan, 291–303, and R. Dawkins's spirited response in Britain to a similar attack from S. Rose, *Nature* 289 (1981): 335 and 528.

90 If it really *is* tentative!

91 'Emic' explanations are those that account for complex social processes entirely in terms of the ideology of the people concerned: that set of concepts regarding relationships and roles that legitimates the behaviour of a people. They amount to causal explanations at the level of the reasons people attribute to their own actions (Crook, *Human Consciousness*, 172).

92 'Etic' observations are explanations by 'disinterested' observers or experimentalists that are dependent upon distinctions judged appropriate by a community of scientific observers (see M. Harris, *The Rise of Anthropological Theory* [New York: Crowell 1968], 576). Etic theory can be falsified through the dismissal of hypotheses by independent observers (Crook, *Human Consciousness*, 173).

93 Crook, *Human Consciousness*, 189–90

94 See, for instance, A.M. Quinton, 'Ethics and the Theory of Evolution,' in I.T. Ramsey ed, *Biology and Personality* (Oxford: Blackwell 1975), 107ff; A.G.N. Flew, *Evolutionary Ethics* (London: Macmillan 1967). For another view of this 'fallacy,' see P. Hefner, 'Is/Ought: A Risky Relationship Between Theology and Science,' in A.R. Peacocke ed, *The Sciences and Theology in the Twentieth Century* (London: Oriel Press, Routledge and Kegan Paul; and Notre Dame, Ind.: Notre Dame Press 1981), 58–78; M. Hesse's comment on Hefner's article in the same volume (pp. 283–4); and the June 1980 (vol. 15, no. 2) issue of *Zygon*.

95 K. Peters, *Zygon* 15 (1980): 213

96 It needs, of course, much amplification along the lines I have developed elsewhere in *Creation and the World of Science*.

97 Re Wilson, see note 73.

98 E.O. Wilson, *On Human Nature* (Cambridge, Mass.: Harvard University Press 1978), 2, 193

99 See note 94.

100 A.J. Dyck, 'Moral Requiredness: Bridging the Gap between "Ought" and "Is" – Part I,' *J. Rel. Ethics* 6 (1978): 293–318; see also, Eileen Barker, 'Value Systems Generated by Biologists,' *Contact* 55 (1976): 2–13.

101 Hefner, 'Is/Ought,' 76

102 In Peacocke ed, *Sciences and Theology*, 28 (see note 90)

103 For further discussion of this issue see P. Hefner, 'Survival as a Human Value,' *Zygon* 15 (1980): 203–12 and W.H. Austin, 'Are Religious Beliefs "Enabling Mechanisms for Survival"?' *Zygon* 15 (1980): 193–201.

104 E.O. Wilson, *On Human Nature*, 201

105 See R.W. Burhoe, 'The Human Prospect and the "Lord of History,"' *Zygon* 10 (1973): 299–375 and D.T. Campbell, 'On the Conflicts Between Biological and Social Evolution and Between Psychology and Moral Tradition,' *Amer. Psychologist* 30 (1975): 1103–26, reprinted in *Zygon* 11 (1976): 167–208. This whole issue of *Zygon* (vol. 11, no. 3, 1976) is a discussion of Campbell's thesis.

106 Job 13:15

107 Hefner's reply to this question would be that 'the most fundamental affirmation in the Judeo-Christian traditions concerning God is that of his faithfulness to and love for his creation ... The theologian has no alternative but to assume God's faithfulness will not allow creation, including the human portion of that creation, to go unconsummated ... When the term "survival" is incorporated within the theological purview, it takes on the meanings associated with consummation and destiny under God ... Christian faith gives the created order very significant status within the purposes of God: if therefore it is determined that the survival thrust is a major motif operative within that order, a motif that gives shape and dynamic to the created order, even where that order includes human beings, the theologian must make the effort to discern how that motif is related to God' (Hefner, 'Survival as a Human Value,' 209–10).

108 See, for instance, the publications of Mary Midgley, E.O. Wilson, J.H. Crook, R.W. Burhoe, and D. Campbell already mentioned and – an important work – G.E. Pugh's *The Biological Origin of Human Values* (New York: Basic Books 1977); J. Jaynes, *The Origin of Consciousness in the Breakdown of the Bicameral Mind* (New York: Houghton Mifflin 1976; Toronto: University of Toronto Press 1978); G. Altner ed, *The Human Creature* (Garden City, NY: Anchor Books, Doubleday 1974); Carl Sagan, *The Dragons of Eden: Speculations on the Evolution of Human Intelligence* (London: Hodder and Stoughton 1977); and H. Harris ed, *Scientific Models and Man* (Oxford: Clarendon Press 1979).

109 A. Hardy, *The Spiritual Nature of Man* (Oxford: Clarendon Press 1979) – an account of the findings of the Religious Experience Research Unit, Oxford.

110 Cf. Isaiah 41:2–4.

111 Temple, *Nature, Man and God*, 130, 131

112 John 1:1–3

Science and the Christian Phylum in Evolutionary Tension

KENNETH E. BOULDING

The word 'phylum' was taken somewhat out of its narrow biological context, and broadened by the Jesuit palaeontologist and philosopher Pierre Teilhard de Chardin. The word 'phylum' derives from the Greek word for 'tribe.' In its narrow sense it means a 'group of organisms having a common ancestor or ancestors.' In Teilhard's sense, however, which is the way it is used in this essay, it means a set of populations of species through time and space that have an important part of their genetic structure in common, derived again from common ancestors. The older the ancestor, of course, the larger the phylum is likely to be. One could think of it as something like a tree. Its roots are the common ancestors, the trunk is the main line of development, the top is the present day, or the time when the phylum peters out, as some of them do, and there are branches, many of which also peter out. The trunk sometimes divides into two or more trunks. At this point the tree analogy begins to break down, and we should not carry it too far.

In biological evolution each significantly identifiable part of the genetic structure of any individual organism could hypothetically be traced back to some date of origin, in a mutation in some ancestor, and even further back to the origins of the structures that mutated. In our own bodies we have structures, for instance, the mitochondria, that may well have been reproducing themselves almost unchanged for over a billion years. That part of our genetic structure which produces the vertebrate skeleton goes back to fairly complex mutations hundreds of millions of years ago. That part which produces the large human brain is probably not much more than a hundred thousand years old. A very small fraction of our genes may be derived from a mutation in one or the other of our parents. Even though the chemical elements in which the structures of our body are coded have very recently been assembled, the patterns that are coded in them are often of very great, indeed, of almost inconceivable antiquity. Parts of all of us exhibit patterns and information structures that may easily be three and a half billion years old.

On the whole, biological evolution has been dominated by what might be called 'biogenetic' patterns. Such patterns involve a sharp distinction between the genotype coded in the genes, mainly owing to the extraordinary molecule DNA, and the phenotype, which is the organism that grows and matures, may reproduce, and eventually dies, and which is produced by the information, or 'know-how,' coded in its genotype, for instance, in a fertilized egg. The genotype is the egg, the phenotype is the chicken. In biogenetic evolution mutation, that is, change in the fundamental information and 'know-how' that produces organisms, takes place entirely in the genotype. Selection takes place mainly in the phenotype, and comes out of constant ecological interaction, as populations of species of all kinds interact in ecosystems. A population is a set of individual elements, each of which is similar enough to all the others that they together make a significant and interesting group, or 'species.' In biological language the word 'species' is often confined to living organisms, but it needs to be extended to cover *all* populations, whether of chemical elements and compounds, physical variables like temperature or velocity, ideas, concepts, and human artefacts like automobiles. All definitions of a species, however, have a touch of arbitrariness in them. In the real world all individuals are unique, and the concept of the species or a population is to some extent a human construct, though it does correspond to certain significant groupings in nature, and we have to remember, modestly, that all important concepts are fuzzy, and that the real world consists of fuzzy sets.

An ecosystem then consists of interacting populations of many kinds of species – atoms, molecules, organisms, genes, humans, human artefacts – combined in soils, rocks, air, plants, animals, people, roads, buildings, gardens, farms, lakes, pollutants, and so on. The interaction may involve crowding, eating and being eaten, fighting (this is rare), breathing, birth and death, and so on. A *habitat* is an ecosystem bounded by some spatial limits, like a pond or coral reef, in which interactions across the boundary are much fewer than they are within it.

The population of any one species will grow if the additions to it are larger than the subtractions. It declines if the subtractions are larger than the additions. This is what I facetiously call the 'bath-tub theorem.' A species has an equilibrium population if there is some level of the population at which the additions equal the subtractions, and the population is constant. Then the species is said to have a 'niche' in the ecosystem. The niche of any one species depends on the sizes of all the other species with which it is in some kind of contact, and which have effects on its additions or subtractions, and therefore its growth or decline. Additions and subtractions may also include in-and-out migrations. In biological ecosystems the existence of a niche for an organism always implies that it participates in the food chain, that it is able to capture the right kinds and amounts of energy and materials, and that it has enough space to grow in and time in which to grow.

Any living object, whether plant or animal, is the result of a process of production from its origin in a seed or fertilized egg. Production is the way we get from the genotype to the phenotype, from the egg to the chicken. The genotype has to be able to capture energy in some form to select materials and move them into the appropriate structures and shapes, and to sustain temperatures in which these various transformations will take place. It must have enough space and enough time. Plants derive energy mainly from the sun and materials from the air and the soils. There are organisms that eat plants, and organisms that eat organisms that eat plants in the food chain. If an ecosystem is to continue in time, it must constantly receive energy from the sun. It must recycle the available materials through the various populations, or find new sources of materials, for instance in the breakdown of rock into soil. An organism that does not fit into such a process in a particular ecosystem and cannot find enough energy or materials, space or time, will not survive, and will not have a niche.

Part of the secret of evolution is that all ecosystems have empty niches – that is, species that would be able to survive and reach equilibrium population in the system *if they existed*. Australia clearly had an empty niche for rabbits in the nineteenth century. Once introduced, they expanded very rapidly, thereby reducing the niches of many indigenous species. Mutation can only be successful if it creates a new species for which an empty niche exists. Most mutations are *not* successful; the variants they produce die out.

Even in biological evolution, a form of evolution takes place that is not strictly 'biogenetic' – DNA and all that. I call this 'noogenetic' evolution.

From Adam and Eve on, the first representatives of *Homo sapiens*, biogenetic evolution in the human race has been very small. The general gene pool of humans seems to have been fairly stable for at least fifty, perhaps a hundred thousand years. There seems to have been little change in brain size or brain capacity, although, of course, there may well be occasional mutations that produce a genetically superior individual. It could also be, however, that genius is simply the result of a lucky arrangement of the genes of the parents and has very little to do with mutation. Adam and Eve (or at least the first reproductive human group), however, already had brains and bodies and capabilities (which were passed on biogenetically to their descendants) that possessed the potential for all that the human race has accomplished by learning in language, art, and science.

Human history is simply the development over time of the extraordinary potential of the human organism for learning, knowledge, and know-how. Because of this capacity, the human race has developed an enormous volume and variety of artefacts. A few pre-human organisms produce artefacts like termite nests, birds' nests, beaver dams, burrows, and the like, but these seem to be mostly the result of instinctual behaviour derived from patterns built into the nervous system by the genes. We cannot wholly rule out the possibility of some noogenetic evolution, even here. Did the beavers, for instance, ever produce a beaver Frank Lloyd Wright, who thought up how to build dams, and transmitted this knowledge to little beavers by a learning process?

A societal evolution, that is, human history, is largely the evolution of human artefacts. These include material artefacts from eoliths and flint arrowheads to the space shuttle. They include organizational artefacts from the first hunting bands to the United Nations, General Motors, and the Catholic church. They also include what might be called personal artefacts; individual persons who possess specialized learned structures – craftsmen, scientists, politicians, and so on. Each human being is in part a biological artefact from the gene structure of the original fertilized egg, and partly a societal artefact produced by inputs of information, both factual and emotional, by which we learn language, patterns of cultural behaviour, and the knowledge and valuational content of our minds. An important part of this information is internally generated, for the human brain has an astonishing capacity for producing information internally in the shape of images, dreams, fantasies, and ideas. An important part of the human learning process is the selection of these internally generated images by a critique that comes through the senses from outside. It is easy for me to imagine that I see a pink elephant, but my eyes tell me that it is not there, unless of course, I am far gone in drink or schizophrenia.

Human artefacts, just like biological artefacts, are created by processes of production that originate in some kind of genetic structure: in this case, noogenetic structures in the human mind that involve 'know-how,' often derived from 'know-what.' If we ask, for instance, why there were no automobiles before 1875, the principal answer is that we did not know how to make them, and even if we had known, before the 1860s there would have been no gasoline to run them. We would give exactly the same answer to the question why there were no human beings ten million years ago – that the genetic structures of the earth did not know how to make us.

Human artefacts are part of the world ecosystem just as much as are biological organisms; automobiles are just as much a species as are horses. They exist because there were niches for them in the ecosystem, and the system knew how to make them. In the case of human artefacts, niches are frequently created by human demands and valuations. Just as biological evolution proceeds by mutations that produce organisms that fill empty niches, so human history consists of mutations that produce human artefacts that fill empty niches in the social ecosystem. Like biological organisms, human artefacts will expand into their empty niches until they reach their equilibrium population at which, therefore, the population will cease to grow. Like biological evolution, human history consists of ecological interaction between human artefacts and the rest of the populations of the planet, which is *selection*, and a condition of constantly changing parameters, new ideas, new discoveries, and new knowledge, which is *mutation*. Just as biological mutation consists of new know-how in a gene, societal mutation is new know-how in the field of human artefacts creating what might be called 'noogenes' – units of learned structure that are then transmitted to each next generation.

There are, however, important differences between societal evolution, which is almost entirely noogenetic, and biological evolution, which is largely biogenetic. In biogenetic evolution, the know-how of the genes is contained in the organisms that this know-how produces. Thus, these organisms can reproduce themselves – in asexual species by cell division, in sexual species by a male fertilizing a female's egg. There is contained within a stallion and a mare together the information that can produce another stallion or mare. An automobile, however, does not contain the information that can produce another automobile. That information is contained in other human artefacts, in blueprints, in libraries, and in learned structures in human brains. Societal evolution is multiparental: an automobile is produced as a result of the coming together of many thousands of different genetic structures from thousands of different persons and other human artefacts.

A societal evolution, or human history, exhibits phyla, for we not only have biogenetic ancestors, we have cultural ancestors. Every word of the English language I speak and write was first used in the sense I intend by somebody in the past, unless it is a word I invented myself, which I can attest to doing sometimes. Each word has been passed on by a learning process from the original inventor, through innumerable intermediaries, to myself at the present day. Just as we could hypothetically identify the date of origin of every structure in our genes, so we could hypothetically identify the date of origin of every structure in our minds. The language that I am using goes back in its present form some four hundred years, developed out of some earlier forms that are still recognizable, going back perhaps five thousand years to the original Indo-Aryan. A few of the ideas that I am expressing I thought of myself, not very long ago. Some of them have come down from Plato or St Augustine, Newton or Darwin, passed on from one mind to the next, generation after generation.

Because of the multiparental nature of personal artefacts, however, the phyla in human history are much more interwoven and complex than they are in the history of biological organisms. The trees of human history are constantly taking strange grafts because of the constant, complex interchange of information among human beings, even in different cultures. Much of the rising European know-how and technology after the fall of Rome came overland from China – such things as the horse-collar, the stirrup, the compass, and perhaps even the idea of printing. European and American technology is now spread all over the world. The early noogenetic ancestors of the automobiles in Botswana mostly lived in Europe and the United States a hundred years ago. Each of the great world religions is a phylum and its origin can usually be identified fairly exactly. Its ideas and practices are transmitted from one generation to the next noogenetically. Science, likewise, is a phylum, its origins perhaps less identifiable and its branches complex.

Two phyla particularly concern us in this essay: the phylum of Christianity and the phylum of science. Christianity originated very clearly in northern Palestine almost two thousand years ago. It can be thought of as a mutation out of Judaism; Jesus and all his disciples were Jews. It had an early graft of Greek or even Roman 'noogenes' as it spread around the Roman Empire, thanks to easy communications and a most remarkable Romanized Jew from Tarsus by the name of Paul. The evidence for the origins is almost entirely in the form of a book called the New Testament, parts of which were almost certainly written within a generation of the death of Jesus, and which seems to have established itself as a 'scripture' within a hundred years. Books are very important noogenes because they mutated very little, even before the age of printing. Like DNA, they can be copied very exactly, although, of course, occasional mutations in the form of mistakes in transcription occur. A 'scripture' is a book of peculiar cultural significance, where the reproduction tends to be very exact, as it is presented, from one generation to the next, as an essential source of the culture it helps to create. A 'scripture' can be compared, perhaps, to the genes that produce the vertebrate skeleton in biological evolution. These are a kind of 'scripture' in the biogenetic structures that replicate themselves over very long periods of time in regard to basic form. The existence of a scripture does not preclude changes in its meaning and interpretation over time as its environment changes, just as the human skeleton is different from that of a fish, even though the basic form is the same.

To vary the metaphor a little, the New Testament can be thought of as the 'oyster shell' of the early church, secreted, as it were, by its life and beliefs and culture. Just as the oyster shell tells us a good deal, though not everything, about the oyster, so the New Testament enables us to reconstruct the life of the organization existing in little cells around the Roman Empire which produced it. The New Testament is somewhat peculiar among world scriptures in this regard. It is not like the Koran or the Book of Mormon, or even perhaps the Bhavagad Gita, which were produced by single individuals, and either written down or dictated by them. The New Testament, as the word 'Gospel' implies, is 'Gossip' about some extraordinary things that happened to some very ordinary people. The gossip was originally oral, and oral traditions can become a kind of scripture as they acquire sanctity. The Gospels, however, were almost certainly collected by the various people who wrote them, from oral statements derived originally from the people who had known Jesus and had participated in the events that are recorded, such as the teachings, miracles, and, of course, the Crucifixion and Resurrection. The Gospels are not philosophy or even theology, but a surviving record of events that later philosophers tried to theorize about and put into systems, such as the idea of the Trinity. The Epistles, of course, were written down from the start, copied, and circulated, and some are undoubtedly earlier than the Gospels and date from the first generation of the early church. The Revelation of St John, of course, is another matter altogether. It is almost certainly very late and it represents the incursion into the church of the stream of apocalyptic thought and poetic diction of which there are one or two traces in the Gospels (for instance, in Mark 13), but which was clearly not the primary interest of the earliest days.

There can be little doubt that the early church was a closely knit community of people who believed that someone who had raised people from the dead was himself risen from the dead. This belief further created immediate charismatic experiences, such as the excitement of Pentecost, and continued with such phenomena as the speaking in tongues, and it is clear that the people in these little communities of the early church found themselves to be transformed by the extraordinary hope that the news of the Resurrection inspired. This hope itself was enough to make a community of the 'saved.'

The church as an institution was perhaps 'saved' by St Paul, who certainly had his own charismatic experience on the road to Damascus. He saw the church as an organization of people in many walks of life rather than as a communal community of the sort that Peter tried to establish in Jerusalem. The church in Jerusalem, perhaps because of its extremely charismatic origins, actually failed. One sees in the story of Ananias and Sapphira (Acts 5:1–11), how a community that started off by having all things in common in a great outpouring of love was invaded by fear.

The history of the church I shall summarize very briefly. It spread throughout the Roman Empire and captured the tottering political structure through Constantine. Whether the church conquered the Roman Empire or whether the Roman Empire conquered the church is a nice point. The church conquered the invaders of the Roman Empire and its successive states and all of northern Europe. It had spread to Russia by the year 1000. It split in 1054 into the western section, based in Rome, and an eastern section based in Constantinople. It had been squeezed earlier by the rise of Islam, finding itself almost in a pincers between the Turks in Asia Minor and the Moors in Spain. It exploded to the Americas after 1492. It tried to penetrate Japan and China in the sixteenth century but largely failed. It expanded to Australasia and to enclaves in Africa and Asia in the nineteenth century. It experienced constant reformations, arising mainly from rediscoveries of the transforming power and belief recorded in the New Testament. Sometimes these took place within the church, as with the Franciscans. Sometimes they were destroyed by the church (the Albigensians). Sometimes they split the church (Luther). Sometimes they created new movements like George Fox and the Quakers or John Wesley and the Methodists.

Like every phylum, Christianity developed subspecies with a wide range of cultural characteristics, ranging from Puritans to 'liberated,' from political to apolitical, from tolerant to intolerant, from peaceable to violent, over a very large spectrum. In each society where the church was important, it was both profoundly influenced by that society and also influenced it. It continues to show vitality in many different parts of the world. Like all things the church has a tendency to decay and become dead and formal. In some places, like Mao's China or Islamic Turkey, it has been almost destroyed by an adverse political climate, but it has shown an astonishing capacity for revitalization – perhaps by the constant rediscovery of the original life that gave rise to the New Testament.

Science is also a phylum with close historical interrelationships with the Christian phylum. Its origins are complex and obscure. It never developed a 'church' although it is beginning to show some signs of such a development in the national academies and scientific associations. It is only in the last two or three hundred years that it has become a world culture. Unlike the world religions, it originates not so much in charismatic leaders and events as in a recurring curiosity about the 'real world' and especially about the material objects and events within it. Pre-Copernican science has a long history, going back to the Greeks, Babylonians, and Hindus, especially in mathematics and astronomy, geography, and a little observational biology. Pre-Copernican science has something of the relation to modern science that Judaism has to Christianity. It existed in small, isolated cultures here and there, passed on through the writings of thinkers like Euclid or Aristotle that survived, together perhaps with some oral tradition. The fact that certain scientific ideas in astronomy and the calendar originated quite independently among the Mayans in Central America and the Anasari of the south-western United States suggests that science is an inherent potentiality of the human race. But conditions clearly have to be right for this potentiality to be realized. An empty niche for it must be ready in the social system.

Copernicus is often regarded as the founder of modern science, although his system is not much more sophisticated than that of Ptolemy from thirteen hundred years earlier. The Copernican solar system with the sun at the centre challenged what might be called 'folk knowledge' that the earth is flat and the sun rises in the east and sets in the west. This new viewpoint perhaps helped to liberate some human minds into that 'testable fantasy' that is the essence of modern science. Fantasy has always been an important activity in the human species. We have the power to imagine things far beyond what we experience. We see this capacity in the pantheons of Greek or Japanese Shinto religion, and in the fantastic epics and folk tales of all cultures. Science, however, develops images that become ever more remote from human experience. The gods, after all, are only humans writ large. Nothing in folk religion is as fantastic as charmed quarks and black holes.

The explosion of science after Copernicus (1473–1543), who was a somewhat younger contemporary of Columbus (1451?–1506) and a slightly older contemporary of Martin Luther (1483–1546), is still somewhat mysterious, but there is not much doubt that it happened. It got under way rather slowly. Galileo (1564–1642) was born twenty-one years after the death of Copernicus. Newton (1642–1737) was born the year that Galileo died. With Newton, however, and the foundation of the Royal Society in 1662 science became part of the establishment, and in England, at least, Charles II might be regarded as the Constantine of science. Economics has some claim to be the third or fourth oldest of the sciences after astronomy and physics, and perhaps geography. Its foundation certainly dates from Adam Smith's *Wealth of Nations*, published in 1776. Chemistry did not really establish itself until Dalton, about 1808; geology with Lyell, about 1830. Biology is a little harder to date. The Swede Linnaeus, who established a taxonomic system of biology, was an older contemporary of Adam Smith. Early evolutionary theory dates from Darwin's *On the Origin of Species* in 1859. Genetics came with Mendel, emerging in the 1870s, but was not much noticed until about 1900. Molecular biology and plate tectonics in geology are products of the last generation. Experimental psychology dates from about 1860, sociology from perhaps 1880. Political science, of course, goes back to the political philosophy of the Greeks, and merged, if ever, into social science so imperceptibly that it is hard to say when.

The impact of science on economic life really only begins about 1860, through the development of the chemical industry, and in the 1880s with the development of the electrical industry. Hybrid corn came in the 1940s, and nuclear power in the 1960s.

The sciences exploded geographically in the nineteenth century to the whole world, first to the Americas, to Japan after 1870, to Australasia and the rest of Asia in the late nineteenth century, and to Africa in the twentieth century. Almost every country in the world now has at least one university where science is taught and in which the periodic table of the elements appears as a 'mandala' on the wall of some class-room.

The interaction between science and Christianity has been very close from the beginning. Modern science arose in Europe, essentially, in a Christian culture. All the early scientists were Christians:

Copernicus and Galileo were Catholics; Newton was a Protestant of a rather secretly Unitarian persuasion; Adam Smith was a Deist; Priestley was a Unitarian; Dalton was a Quaker. It would be an interesting question for historical research to determine who was the first atheist scientist. It may well have been Laplace of whom the famous story is told (the origin of which I have been quite unable to trace) that he replied to Napoleon, who asked if he found God in his system of equations of the solar system, 'I have no need for that hypothesis.' Science is often associated with the Enlightenment of the middle eighteenth century, which produced the American Revolution. But the leaders of the Enlightenment, such as Voltaire and Rousseau, Thomas Paine, and Thomas Jefferson, were not primarily scientists. The Enlightenment was a movement among philosophers, essayists, and politicians, rather than scientists.

A very interesting question is why science developed in Europe and not elsewhere. Why, for instance, did it not develop in China? This is sometimes called the 'Needham' problem, after Joseph Needham, who has devoted his life to studying this question and has produced some twelve volumes or more of the history of science and technology in China. The Chinese certainly led the world technologically for almost two thousand years. In about the year 1000 one would have seen China as indeed the middle kingdom, with the highest technology – surrounded with a crescent of Islam almost as developed and preserving the heritage of the Greeks, which had been nearly forgotten in Europe – and Europe would have looked like a little obscure peninsula on the edge of the civilized world, just emerging from the Dark Ages.

Nevertheless, some main reasons why science did find an empty niche in European society and not elsewhere may be sought in examining the various religious backgrounds. Christianity was a religion of working-class origin, and consequently much closer to the material world and its reality than the philosophies of Brahmin India or Mandarin China. One of the main differences between Buddhism and Christianity is that Buddha was a prince and Jesus was a carpenter. It is significant that Peter was a fisherman, Paul was a tent maker, and

Jesus was born in a stable. A society like India in which the prevailing doctrine, at least among the intellectual classes, is that the material world is *maya*, an illusion, can hardly be expected to produce much in the way of physics and chemistry. Curiously enough, the Christian doctrines of the Trinity, the Holy Spirit, and the Incarnation also emphasize the manifestation rather than the intervention or intrusion of God in the material world.

I am not familiar enough with the subtleties of Islamic doctrine to do more than speculate whether the strictly monotheistic character of Islam separated God from nature, and hence pushed Islamic culture into ritualistic religious practice, especially in petitionary prayer, rather than into the investigation of nature as the manifestation of God. It may well have been, however, that Islam was just on the edge of science about 1200, especially in Baghdad, and that what prevented the rise of science in Islam, and perhaps forced it to be almost a fossil culture for so long, was the destruction of Baghdad by the Mongols and the slaughter of most of the Islamic intelligentsia.

There are, no doubt, other reasons also why science originated in Europe. Europe had what might be called an optimal degree of political disorganization and decentralization, and so failed to discourage innovation in thought. China, perhaps, was too well organized. There was no separation of church and state in China. The mandarin was the priest in Confucianism, and Taoism, closer perhaps to the working people, never found a 'Constantine' and so only had sporadic influence on the political establishment. In Europe, however, the separation of church and state goes right back to the emperor and the pope, and becomes actually a little less noticeable in Lutheranism, as Lutheran churches became state churches. The new ideas of science then could survive in the cracks of the system. Copernicus got away with it in Poland, perhaps because he was so far from Rome. Bruno did not get away with it in Rome, and Galileo was reported to have gotten away with it only under his breath. Tycho Brahe, however, a contemporary of Bruno, did get away with it in Denmark. It is the business of organized political structures to defend themselves against change, and to prevent the opening of any empty niches. It is only as political structures of church and state, business and state, and science and state can be played off against each other that advances can be made.

In spite of, or perhaps because of, the fact that modern science can be quite properly regarded as a cultural mutation out of European Christian culture, the mutual tensions between science and Christianity have been acute and the confrontation is by no means over, as the political activity of creationist groups in the United States indicates. Part of the problem is that the doctrines of Christianity were stabilized in an essentially pre-scientific era, so that the world-view and much of the metaphorical language of the Bible are pre-scientific. It is not surprising, therefore, that the church was so upset by Bruno and Galileo, for the heliocentric view of the solar system and, still more, the vastness of the astronomical universe, as it began to unfold, upset the traditional geography and cosmography of the church as we see it, for instance, in Dante's *Inferno*. The medieval church ruled perhaps more by the fear of hell than by the love of God, and the cosy three-storeyed universe – with hell underground and volcanoes as its vents, earth at ground level, and heaven above the sky – was an easy cosmology for folk belief. I don't know whether Dante really believed that purgatory was at the South Pole, but nobody at his time had ever been there, so it was at least a plausible place to put it. In the cosy cosmology of the Christian Middle Ages, God in the sky was not very far away, and heaven and hell did seem geographically quite real. Even in my Methodist Sunday school, as a little boy, I remember the teacher singing, 'O, see the sky, so blue, so high, so very far away; Who lives up there where all is fair, dear children, can you say?' to which we would reply, 'God lives up there, where all is fair and blue and high and bright. So great, so kind, none can define, He gives us day and night.' When the planets are moved not by angels or by the majestic hand of God, but by differential equations of the second degree, the astronomer clearly has 'no need for that hypothesis.'

What Copernicus and Galileo did for the medieval system of cosmology, Darwin did for the biblical view of time. Not only did the cosy six thousand years of biblical history expand into billions of years of evolution, but creation was made unnecessary, or at least pushed back into a misty past, a very long time ago. Actually, the Big Bang theory of modern astronomy looks a little bit more like Genesis than the rather indefinite evolutionary process of Darwin. Nevertheless, there is still the doctrine of natural selection undermining creationism in regard to the material world – and in particular, of course, living organisms. It is no wonder that Mrs Darwin, who was a good Episcopalian, was rather worried about her husband's immortal soul.

Freud belongs to that uncertain frontier between the humanities and the sciences. Still, by exorcising guilt by reference to childhood trauma, he certainly seems to undermine the Christian plan of salvation and, of course, he himself regarded religion as an illusion.

Molecular biology hardly improved matters by attributing the growth of all living organisms to the know-how contained in DNA and the gene, thereby eliminating even further the necessity of a spiritual creator. The recent theory of autopoiesis makes it even worse for creationism by suggesting that organized structures develop by 'self-creation,' through the way in which random events change the ambience around them when they happen, so that all order emerges in the universe, not by conscious plan or design, but simply because of the inherent instability of chaos.

These scientific images are hard to reconcile with the religious experiences of prayer, meditation, worship, transcendence, salvation, healing, even miracles. It is ironic, however, that in what is supposed to be an age of science it is fundamentalist religion that seems to be winning out politically: in Khomeini's Iran, Falwell's United States, Orthodox Romania, or even the Pope's Poland. Even in Russia, the Orthodox church and the Baptists seem to be much livelier than the spiritually dead and oppressive Communist party. In the great ecosystem of the mind, it may well be that science is like the beasts of the field, grunting away and snuffling at the earth, whereas religion is the birds of the air, carolling delectably, safe from attack. Certainly, science these days cannot take its political legitimacy for granted, and it seems to be threatened by various ideologies and theologies which the powers that be see as a support for their legitimacy. There are certain parallels between the Lysenko episodes in the Soviet Union, the Scopes Trial in Tennessee, and the current creationist and anti-abortionist movements.

Is there then any opportunity, or even any need, for a rapprochement between the world-views and life experiences that are characteristic of science and religion, especially of Christianity? It could be argued, indeed, that there is no need for such a rapprochement. What we have is coexistence and this indeed is the rule of ecosystems, which tend to put a low survival value on uniformity and consistency, and a high survival value on diversity. The optimum number of species in an ecosystem seems to be quite large. This apparent fact is one reason for worrying about monoculture in agriculture, like the Irish potato. Monoculture of the mind is equally dangerous, as the disastrous effect of the Spanish Inquisition in Spain, Hitler in Germany, and Stalin in the Soviet Union certainly indicates. Maybe the beasts of the field and the birds of the air should be happy to live within their own sphere and not try to talk to each other.

Humans, however, are *both* 'beasts of the field' and 'birds of the air' in this day of airplanes, especially gossamer airplanes, and somehow we seem to want more than coexistence, especially in so far as both science and religion have claims to be interested in truth. The feeling that truth must be one, or at least a consistent, structure is hard to overcome. Having lived in two worlds myself most of my life, I can testify that the experience is exciting and also a little uncomfortable. The discomfort is never expressed better than by Tennyson's *In Memoriam*:

> The wish, that of the living whole
> No life may fail beyond the grave,
> Derives it not from what we have
> The likest God within the soul?

Are God and Nature then at strife,
That Nature lends such evil dreams?
So careful of the type she seems,
So careless of the single life,

That I, considering everywhere
Her secret meaning in her deeds,
And finding that of fifty seeds
She often brings but one to bear,

I falter where I firmly trod,
And falling with my weight of cares
Upon the great world's altar-stairs
That slope thro' darkness up to God,

I stretch lame hands of faith, and grope,
And gather dust and chaff, and call
To what I feel is Lord of all,
And faintly trust the larger hope.

So careful of the type? but no.
From scarped cliff and quarried stone
She cries, 'A thousand types are gone;
I care for nothing, all shall go.

'Thou makest thine appeal to me.
I bring to life, I bring to death;
The spirit does but mean the breath:
I know no more,' And he, shall he,

Man, her last work, who seem'd so fair,
Such splendid purpose in his eyes,
Who roll'd the psalm to wintry skies,
Who built him fanes of fruitless prayer,

Who trusted God was love indeed
And love Creation's final law –
Tho' Nature, red in tooth and claw
With ravine, shriek'd against his creed –

Who loved, who suffer'd countless ills,
Who battled for the True, the Just,
Be blown about the desert dust,
Or seal'd within the iron hills?

No more? A monster then, a dream,
A discord. Dragons of the prime,
That tare each other in their slime,
Were mellow music match'd with him.

O life as futile, then, as frail!
O for thy voice to soothe and bless!
What hope of answer, or redress?
Behind the veil, behind the veil.

What might be called the evolutionary '*angst*' has never, perhaps, been better expressed than by Tennyson. An extraordinary thing is that this poem was written probably more than a decade before Darwin got the idea of natural selection through reading Malthus and a whole generation before the publication of *On the Origin of Species*. Perhaps it is only the poets who know the future! It is hard to find any corresponding quotation for the scientific side of the story. It is significant that science has inspired so little in the way of poetry, music, or art by comparison with religion, and that it has been even more subservient to the demands of a militaristic state than perhaps even the church. The Bible gave us Handel's *Messiah* and Bach's *Mass in B Minor*. Science gave us Hiroshima and Nagasaki, and maybe much worse things to come.

If any rapprochement can take place, it must be through mutual modesty. Modesty in the form of 'we may be wrong, but this is how it looks now' is an essential element of the ethic of science. It is of the essence of scientific theories that they are capable of disproof. Many of them have been disproved, and many more will be disproved. This is particularly true of the historical sciences, such as palaeontology and evolutionary biology, for all we know about the past is the record of it. The record is not only extraordinarily incomplete but also highly biased by durability. Human history and all historical records, including the scriptures, are subject to the same limitations.

The experimental sciences may be a little more cocky as they live more in the 'here and now' and can create their own records. Even here, however, a certain modesty is in order. What the experimental scientist cannot deal with is genuinely rare events. It is only the reproducible – that is, the commonplace – that can be the subject of a laboratory investigation, and experimental methods can only deal with events of very high probability within their framework. Such sciences can only study events of low probability if we have a large enough universe to make them highly probable in the mass – like the present search for the disassociation of a proton in the middle of huge quantities of water. Events, however, that can only happen once in a million years are extremely hard to study in the laboratory. Science could therefore develop a more decent agnosticism about its own view of the world than it now has. This attitude would be entirely consistent with the inner logic of science itself.

The creationists' pressures may have actually improved the biology textbooks, making them a little more modest, saying that things *may* have happened this way but we are not quite sure that they did. A similar modesty is hard to find in fundamentalist religion, but would be equally becoming. Modesty would suggest that religious language is largely metaphor. There is nothing wrong with that, for it is dealing with ultimates and incomprehensibles. It is immodest to suppose that our finite minds can fully comprehend all that there is to be known in this very vast universe. I have much sympathy with the ancient Hebrew taboo on the use of the name of God, and modesty demands that we should not be over-confident or chatty about the large design of things. In true prayer or worship we reach out and are met; in the mystery of Communion we are fed. This is a fact of human life and experience that can be described, but cannot be explained. In humility and modesty we can only adore. Perhaps one day even the beasts will fly.

Some Aspects of Scientific Creativity

P.W. KENT

It is recalled (Monod 1974) that when Einstein came first to Paris a well-known hostess arranged a soirée at which he and the poet Valéry, great as a poet if perhaps less so as a philosopher, met. Valéry, who had a great interest in the process of creativity rather than in the products of creation, began to ask Einstein about his work habits.

'How do you work, and could you tell us something of this?' he asked, amid a group of curious hearers.

Einstein was vague. 'Well, I don't know ... I go out in the morning and take a walk.'

'Oh,' said Valéry. 'Interesting, and of course you have a notebook and whenever you have an idea, you write it in your notebook?'

'Oh,' said Einstein. 'No, I don't.'

'Indeed, you don't?'

'Well, you know, an idea is so rare.'

This exchange in itself tells us a good deal about really great creativity. It identifies some of the ingredients that are common to the background of creative thought – leisure (solitude) and preparedness for a dynamic mental process to recognize and identify what Einstein calls his 'idea.' We know from later comments that this identification, that is, specifying that which is worthy of testing, is a dynamic. Einstein confesses to many thoughts teeming through his mind which, by mental sifting, result, on rare occasions, in a thought being retained for further critical thought and reflection, on which he possibly confers the status of an idea. In 1914, two years before the General Theory of Relativity, Einstein is said to have had the outline of an idea once every two minutes, the great majority at once rejected. 'Our whole problem is to make our mistakes as fast as possible.'

PREDISPOSING CIRCUMSTANCES

Imagination is one of the foremost ingredients in the creative process, though one suspects very different meanings and nuances are attached to it by different writers. Whatever the philosophical implications, it has to be recognized that for the working scientist there are substantial difficulties, and a diversity of professional opinions, about the concept of imagination and its relation to the area designated 'belief.' A scientist in his world inevitably thinks of images as symbols (mathematical, atomic, conceptual, and so on) connoting structures and subsuming their accompanying properties. Imagination, conscious or subconscious, plays games with such symbols and, at times – intuitively, we say – a strikingly unusual correlation arises, if the individual is sufficiently alerted for it to register. One recognizes that in the sifting process correlations – ideas – possibly of considerable importance, can arise which we, the majority of ordinary mortals, fail to recognize and so pass by. For the genius, perhaps, matters may be different, with a heightened ability for symbolic correlation and apprehension of what is important and worthy of further study. 'Genius is to see what everyone has seen and to think what no one has thought.'

In the scientific context, ethical descriptions do not become attached in the first instance to ideas – whether they are good or bad – though subjectively many people without thinking regularly apply such quality judgments. A 'good' idea, Einstein's rare idea that is, generally may be no more than one for which no particular falsification comes to mind and which thus merits further investigation. The subsequent detailed critical scrutiny of the idea and its possible acceptance into the consistent body of knowledge is as far as we can take the matter.

In recent years a different aspect of ethical consideration has come to be applied to certain areas of scientific investigation where there may be grounds for believing that serious socially damaging consequences could be involved. The notion of the responsibility of scientists, emerging as it did originally in connection with genetic engineering, that is, replication in bacteria of viruses pathogenic to humans, has gained wide acceptance. Codes of practice governing experimentation in this and other areas have been embodied in statute law in a considerable number of countries. This thinking extends more broadly to the selective use of experimental animals and to a generally greater caution in drug design and testing. Following wide-ranging international exchanges between biological scientists, agreements are forged regarding those types of experiments that are considered proscribed – that is, that become *bad applications* – though speculation may well continue. But at the same time and of equal importance, public opinion has become aroused and alerted to take a markedly more cautious stance towards the claims of science. The doubts of the conservationists and environmentalists have been reinforced.

To attempt to define general predisposing circumstances for the generation of an idea is not feasible if it is admitted that such an event is not only rare but is also peculiarly personal. We can, however, examine some sets of broadly predisposing circumstances from the statements that distinguished innovators have made about themselves.

One such set may be illustrated by Jacques Monod who in asking himself 'What is discovery in science?' begins by examining the logic of the process in the view that there is no creativity without logic. In 1974 he wrote, 'Our task is simple because all we have to do is to abide by Popper's Principle of Demarcation.' 'If I may oversimplify scientific discovery consists of stating assumptions based on the hypothesis that their structure may be falsified even by an imaginary experiment.' It would seem to follow that there can be no such thing as a purely empirical discovery; there can only be an assumption that precedes the empirical basis of 'fact.' The Monod position regards discovery as a deliberate logical process in which the discoverer sets himself in a particular relation with regard to his field of interest. But it may be argued that this process belongs to the second aspect of 'application,' or 'testing' the prior idea. To adopt a predictive attitude, even without so realizing, the investigators identify themselves subjectively, in some measure, with the situation. How would an electron or an enzyme behave under this or another set of experimental conditions? The investigator is open to be influenced, no matter how minimally, subjectively as well as objectively, to engage in a systematic and selective search for mathematical or experimental results. Monod's own research on oligomeric enzymes arose from kinetic studies of known enzymes that failed to conform to accepted theory. The deductions which followed falsified accepted theory and required it to be rewritten. This type of continuous revision, recasting, and refining of accepted theory forms a fundamental part of scientific thought and marks it with a transient quality. As in any other form of discovery, the necessity for a common language and common definition for the formulation of symbols constitutes other fundamental ingredients. Since we are thinking of scientific creativity, we can take experimental technique into the concept of language. What is held to be meaningful and what may be 'falsified' is dependent on the accuracy and appropriateness of the techniques available. It is by no means uncommon in the various theoretical natural sciences for interesting and impressive ideas to be unavailing through lack of suitable experimental or mathematical techniques sufficiently specific or exact for the Popper Principle to be applied. Such ideas may be recorded and set aside or may simply be rethought at some later time when they become examinable with new techniques. In a comparable way experimental data accumulate, adding to a pool of derivative secondary considerations, until an idea produces fresh primary correlations. In current science this pool of collected thought forms a further essential feature of the background in which creativity arises.

A second set of circumstances accords with the experiences of other distinguished scientists (for instance, Krebs, Eigen, Eccles) who, within the mastery of their specialties – language and techniques – *care* for their subject and in a sense become haunted by it. Here, 'good ideas' are spontaneous, arising unexpectedly and irrationally (though not necessarily illogically). 'Intelligence,' it has been commented, 'is not enough ... what matters is the willingness to make proper use of intelligence.' The apparent subconscious ability to select – intuition – is found in every area of innovative science of every degree of sophistication, whether in highly abstract fundamental theory or in issues of relatively simple content – for instance, Fleming's observations leading to the discovery of the penicillins or the elegantly simple invention of paper chromatography as an analytical tool.

The description of this unconscious activity can be further illustrated by the well-documented story of Otto Loewi, who during one night in 1916 had an idea about the transmission of nervous impulses. He jotted down a few notes, but when he awoke in the morning, alas, he could not decipher his writing nor was he able to recall the idea. Remarkably, the idea returned to him during the following night. This time he got up at 3:00 A.M., went to the laboratory, and did the experiment. It was successful, and fifteen years later brought him the Nobel Prize. If one is realistic, it is difficult to avoid the conclusion that in the generation of ideas there is still a measure of that earthly quality of chance usually called luck. 'Chance favours the prepared mind,' as Pasteur put it. For though we strive to understand such processes in rational terms, there is always a residue that is not rational. At best, our rational understanding of much of the world in general and of human behaviour in particular is still wildly incomplete, and such scientific correlations that are made are thus under the shadow of our ignorance.

Chance (or luck) of course can act both ways, as it did for Sir John Eccles who tells that he too, like Otto Loewi, had a dream, following a period of intense thinking about processes within the central nervous system. He dreamed how inhibition might work. In the morning he remembered his thoughts and performed the experiments that resulted in the construction of the theory that was subsequently published. Unfortunately it was wrong, and was not refuted until three years had elapsed, though later it was shown to be valid in a number of special cases. Thus it ended, as do so many scientific discoveries, found to be incomplete (or wrong), displaced, or updated, and then reinstated in some new form.

In both sets of circumstances – Monod's systematic approach and Loewi's intuition – each individual is in a high state of imaginative stimulation, one may even say fantasy, not arising solely from here-and-now immediate sensory input. This type of fantasy is concerned with the 'reality' of language, of techniques in a positive sense in contrast to the negative Freudian fantasy of evasion. In the free-ranging use of imagination, the creative scientist resembles the artist in that elegance, simplicity, and symmetry, if not beauty, may be displayed. It is in the application, that is, consequential refinements of techniques and exemplification, that the two approaches are most clearly distinguishable, but for either to be successful, each has to have some qualities of dreaming. Imagination, conscious or subconscious, is a prerequisite for all stages of creative thought conducted on more than a trial-and-error basis. The human contacts by which imagination is stimulated are not to be underestimated, and the day-to-day contacts between fellow-scientists sharing common interests have powerful effects. This aspect of the creative process is particularly relevant to an education of young scientists in the ways of understanding, in contrast to a training in the absorption of established information.

Among further prerequisites for productive creativity, motivation plays a dominant role, and this concept brings us to further considerations.

IMPLEMENTATION OF CREATIVE THOUGHT

In addition to the circumstances which predispose to originality, the range of human, subjective, and social qualities is identifiable. The urge to give vent to originality rarely stems from intellectual pressure alone, and the satisfaction in sharing a good idea with others, particularly if it contains the solution to some vexatious problem, can be a source of intense personal happiness. Personal achievement follows too from the successful outcome of the intense and lengthy periods of hard work that painstaking experimentation can entail. In meeting the challenge of being creative, the chances of disappointment are great and frequently encountered, but the possibilities of satisfaction in overcoming the challenge are felt to make it worth while.

As a human activity, other elements are involved in the motivation of scientific creativity and it has to be recognized that acquisitiveness, *sub rosa*, is one such, and pervasive, element. The high esteem given by scientific opinion to the innovator in the form of priority of publication – 'the urge to be first' – creates not only motivation but a climate of competitiveness that may be neither healthy nor friendly. In some sectors, including much of the academic world, chances of promotion to lucrative or influential professional posts are all too often thought to be enhanced by the individual's rate of publication. The hazards of being original to order becomes a regrettable and debasing factor, fostering motivation of an undesirable form. Though further instances of other diversionary trends could be cited, it is perhaps a happier thought that financial acquisitiveness appears to weigh less as a motivating factor among scientists.

The formative years of the creative scientist play a crucial role in generating a sense of priorities and values among these various ingredients. It is impressive and touching to note the number among the world's outstanding scientists who acknowledge, with gratitude, the influences of those who taught them. The history of scientific creativity repeatedly confirms that distinguished pupils were the products of distinguished teachers through personal contact and example. Most will be familiar with the academic 'family-trees' showing the 'parentage' of many of our best-known scientists, and there is no doubt that these relationships have been of great importance, especially in this century, which has seen a phenomenal growth in scientific endeavours of all kinds. In the 1930s it was claimed that, as far as inventiveness was concerned, 90 per cent of the progress was contributed by 10 per cent of the scientists, while by 1973 it is said that (in physics) 10 per cent of the physicists accounted for 50 per cent of the progress. It seems fair to assert that there are few architects and many workmen (with an intermediate group of interior designers). That excellence begets excellence is a view popular, on the whole, with those in government and in the counsels of bodies whose duty it is to fund and promote research.

But even given motivation, dedication, and unremitting commitment, creativity does not automatically follow. A great deal of labour, passing for research effort, can be dissipated on experimental work destined to be unproductive through faulty design or lack of the 'good idea.' The temptation of the enthusiastic researcher, senior as well as junior, to overwork, even to the point of ill health, is by no means uncommon. Although for brief interludes it may be necessary, such a persistent practice can be mischievous, if not damaging. An idealist approach, attributed to a well-known biochemist, is that of performing the absolute minimum of experiments, but each one exquisitely thought out and designed in strategy and techniques. All too often researchers are content to conduct repetitive investigations on a lower plane – an activity akin to data collecting and original only in the sense that the information has not been collected before. As has been mentioned, such data have their value in contributing to the general background out of which new fundamental correlations arise.

In contrast to this complacent attitude, many creative scientists possess a pervading dissatisfaction, if only with existing hypotheses. Their motivation contains an urge to renew and refine. This element of nonconformity is creatively productive though clearly it cannot be taken to the extreme or, as has been said, we would have to start where Adam started and there is little reason to suppose that we should get much further than he did. Qualified nonconformity may require a measure of technical courage as far as the experimentalist is concerned. It may be difficult for one brought up with, and having acquired command of, one type of methodology to relinquish it and adopt another. But the senescence of techniques can swiftly become a limiting factor in the creativity pattern. Only if techniques, like hypotheses, become refined and are made precise and specific, can they become capable of testing fresh ranges of new ideas.

INCULCATION OF CREATIVITY

Some mention has already been made of the educational background within which creative people of science seem to have flourished. But it is interesting to speculate whether it is possible in any way constructively to inculcate creativity. One is aware of course of that favourite of liberal arts colleges, 'creative writing.' In 1967, a provoking survey (Pickering 1967) was made of the role of universities, especially those in the United Kingdom, in contributing to the renewal and stimulation of national cultural life as well as the promotion of material prosperity. The well-being of society was seen as being at risk through contemporary complacency and stultification brought about by inadequate encouragement of young talent and by a pervasive quest for egalitarianism. Pickering's argument was based on the rarity of really creative minds with an innate capacity for intuition and ideas – that is, the innovators. In an age become closely dependent on its scientific development, the future had become dependent upon such minds. The case, undoubtedly élitist (a much misunderstood description), has much self-evident authenticity and would be fully convincing were natural sciences the key to all contemporary problems or even to some of the most pressing ones.

The over-simplification is that such is not the case and that the most advanced technological societies are not, nor have been, consistently the most harmonious or effective. The question can be posed: how was it that in the Depression of the 1920s and 1930s the most technically developed countries fared worst? Even within a decade of Pickering's book, institutions of higher education in the West, fully conscious of their role in scientific education, were in assorted states of disarray and turbulence through influences remote from the world of the physical sciences. Indeed, a trend that emerged from that phase, and gained ground, was a mistrust of science and the scientist which, as will be discussed later in this paper, was to have a deep and lasting effect.

The focusing of attention on universities and similar institutions can lead to a view that these provide the sole road for the creative scientist to tread. For many, this conclusion may be correct because of the universities' general circumstances of providing an informing environment in personal qualities (self-denial, dedication, persistence, and so forth). But these are also the general attributes in which the creative endeavour can rise to the surface, and the detection of the innovators may well be one of the most significant functions of current higher educational science (Carter 1980).

But the university scene does not have the prescriptive title to uniqueness nor, to be fair, would it claim to. Throughout recent history some of the most illustrious have risen outside the walls of academe, especially in the world of affairs (Abraham Lincoln, Churchill), or have made their major contributions in the wilderness, as it were. But even in the last century the wilderness was suspect. The announcement by van't Hoff and associates of the tetrahedral, three-dimensional arrangement of the carbon atom which was to establish organic chemistry was bitterly criticized by those within the scientific establishment. In 1877 Kolbe wrote: 'It is indicative of the present day, in which critics are few and hated, that two practically unknown chemists, one from a veterinary school and the other from an agricultural institute, judge with such assurance the most important problems of chemistry, which may well never be solved – in particular, the question of the spatial arrangement of atoms – and undertake their answer with such courage as to astonish the real scientists.'

Educationalists thus have to consider the width of the general circumstances in which creativity can assert itself and speculate whether one can do more than foster the special resources for it to flourish.

WIDER ISSUES OF CREATIVITY: SCIENCE AND RELIGION

In the context of the discussion, so far, attention has been largely confined to intellectual and personal attributes involved in the creative process. But the literature of the past century and more, dealing with the relevance of science, abounds with strongly held views about the spiritual element in human nature (which few would deny) and its relation to organized religion, Christianity in particular (which many would). The very persistence of the debate between science and religion, or theology to be more exact, may well seem to keep alive issues that have in fact become moribund. It is not that the horse has bolted, leaving an empty stable, but rather that it refuses to go, and worse, seems to be capable of changing its spots. So long has the debate been with us that it, itself, may well be elevated to the status of myth (Peacocke 1979). Within the community of scientists there is discernible a prevalent attitude of caution, a reluctance to extend scientific claims beyond their pertinent boundaries. A minority, and at times quite a small minority, has taken a different stance in promoting a scientific triumphalism. Though fully aware of the limitation of the scientific view, the pressure of its enquiry is argued to be such that it must eventually give the complete interpretation to the whole of life; the 'final solution is just around the corner.' Thus the debate has been conducted as a conflict of disciplines – sets of propositions, reinforced on the scientific side by experimental techniques and by practical applications to everyday life. At no time can the arguments be represented as between fixed positions. The systematic, continuing enquiry represented by scientific research provides a changing, fluid body of understanding about certain aspects of creation, about man and about life, within the boundaries of the premises, techniques, and attitudes of its practitioners. The theological involvement, no less, provides arrays of insights about the nature of man and the meaning of life, its fulfilment and purpose, again within techniques and limitations of human capability.

The debate may be viewed not only as between intellectual disciplines but as between people – scientists and theologians – and much of the heat has been generated from the subjective extrapolations that individual participants have experienced, often with deep passion. The 'uneasy truce between science and theology' has not been concerned so much with the hardware as with the whole personal outlooks engineered by individuals and built into the background of the arguments. There have been some on both sides destined to hold their ground no matter what new armaments appear on the field, and even these individuals may experience a form of motivation akin to wishful thinking that, misguided or not, belongs to the realm of the imagination. Those of a cautious nature are generally content to compartmentalize their activities and in their one mind seek to keep apart their scientific experiences from their religious and psychological activities of everyday life, with integration at those rare points of contact such as ethical considerations provide. Others, aware of the dynamic state of scientific interpretations as well as of theology, explore ways of achieving, at least for their own personal peace, further kinds of reconciliation.

The growth of public concern – that is, by non-specialists – is of interest and importance, susceptible as it is to the more sensational but often less representative interpretations. The consequent stirring and re-stirring of the popular imagination has given rise to the upsurge in science fiction in the media and in literature. Behind this development would seem to be a love–hate attitude by the general public towards modern science and the scientist. The ready and unquestioning acceptance of the applications of science in the past has gone along with the common assumption that human problems of the future will be solvable by science as we currently perceive it. With this belief has come the corresponding decline, in developed Western countries, of popular interest in theological learning and values. In practice this change has meant that for masses of ordinary people the extent to which life has become longer, more comfortable, and relatively pain-free in the second half of the twentieth century has diminished the urgency to contemplate ultimate questions. It has become no longer acceptable to talk about death or evil, moral responsibility or conscience, or, indeed, virtuous living. Popular imagination in part has come to impart a beneficence to science and technology which would come to the rescue eventually. In relatively recent times, however, there has emerged greater evidence of popular mistrust, or even fear, of science and technology which, as will be considered later, substantially alters the future.

DYNAMIC DEVELOPMENT

It is a general property of creative scientific endeavour that it submits itself to constant and rigorous refinement. Save for the briefest moment in modern times, if ever, could it be said that science occupied a static, 'set-piece' position. The scientific insight available to those in the last century who joined in the science–theology debate was by present standards rudimentary, even for the greatest leaders of the times. Now every high school leaver can have a vastly more sophisticated knowledge of, say, biological sciences than was available to Charles Darwin. It is disquieting when one reads, therefore, of the intensity with which those scientific giants of bygone times contended for points of view long relegated to the lumber-room. This attitude of mind was not the prerogative of the scientist, and at the same time a comparable fixity was present on the theological side, as instanced by fundamentalist contenders. Neither helped interdisciplinary debate in any fruitful direction. Whether the pace of scientific discovery will continue at its present rate is a matter for speculation. It seems likely to do so, or possibly in an even more accelerated way owing to the automation of data-gathering and processing. In astrophysics, for example, E.J. Kibblewhite notes (1982) that a modern telescope can record 10^6 astronomical images in a single hour (about a half of which are stars and the remainder extragalactic objects). Each photographic record contains some 4×10^9 bytes of information so that on a single night, with one telescope, 40 billion bytes of information can be collected. In the processing of such data (even given automation), sorting out random errors, and deciding what could be 'useful,' the human factor may be paramount or, in the end, limiting. Comparable accelerated data acquisition exists in other areas, as for example in nucleic acid sequencing (a task which twenty years ago would have been formidable and which now is virtually a routine matter occupying days rather than weeks). The growth of genetics and immunology correspondingly has yielded results of bewildering complexity.

One cannot but assent that, whatever future pace science may entail, the most modern scientific views accepted today will have been drastically modified and embellished by the end of the twentieth century. By that time, what we now hold true will almost certainly have become simplistic or possibly even humorous and be regarded as highly incomplete. Only by acceptance, in an attitude of humility, of the transience of scientific insights is their true significance and correlation likely to be discerned. Scientific research not only produces pieces of new information leading to new understandings, but it also reveals, in the process, the vastness of our ignorance.

A CENTURY OF CONFIDENCE

The new impetus which the science–theology debate took following Darwin's publication on evolution in 1859 had a profound impact both on specialist and popular opinions. The initial challenge to the then accepted Christian belief was to be argued as much or more on the extrapolations of the evolutionary hypothesis than on basic scientific theory. The prime propounder of the theory for the most part took a most cautious line, aware of the tenuous scientific evidence for the hypothesis. The rapidity with which public interest was awakened by this new impetus, accompanied by the creation of national and personal wealth which technology had brought, not surprisingly fostered a new materialistic climate of confidence in science. It was perhaps this overoptimism that made the claims of religion seem less significant. It was not that Christianity was not true, it just seemed to be irrelevant. Life could be lived within a shell of well-being, if the individual was lucky enough not to be disturbed by fundamental questions of purpose and destiny.

Such traits carried over in even more developed forms into the twentieth century and are evident among a variety of scientific authors. The trait of scientific triumphalism continues in various forms. The reductionist contention that all life is explicable in terms of the laws of physics and chemistry, or in the structure of DNA, has created considerable discussion. Other scientists could assert dogmatically that in their specialist (professional) view no divine principle could be discerned. But these assertions are detached from other aspects of individual experience, sometimes called personal knowledge (Polanyi 1959), which arise from the spiritual dimension and which are present in even the most determined materialist. It is assumed, however, a priori that no divine principle is capable of detection or that it would be meaningless to consider such questions or that the spiritual element in human nature is irrelevant to inspiration, intuition, imagination, or altruism; then no progress is possible.

The mood of confidence in scientific beneficence was to receive a profound shaking first in the world war of 1914–18 when the horrors of misapplied technology on a huge scale inescapably were revealed to all. That such misuse of science should have occurred has been interpreted in terms of a lack of insight, by either governments or politicians, or of errors of the socio-economic order. Doubt, further reinforced in the Second World War, continued to gain ground in the post-war years, and the year 1969 is of particular interest. In that year, man first reached the moon, and for the first time was able to look back at earth and be starkly aware of its finitude and limitations. At the same time, there was a growing awareness in the general public for the need to have respect for life and for the environment. The conservationist movement, extending such thoughts further, pressed for greater regard for the use of earth resources, the problems of an expanding world population, and the need to elevate human values over the materialistic.

Behind these trends was a growing and deep-seated (if not entirely justified) doubt, possibly even fear, of modern science. The effluents of industry, the flowering of technology, were held accountable for grievous pollution, the loss of natural amenities, and violence done to wildlife. Tragedies, such as the thalidomide disaster, were laid at the door not only of scientific exploitation, but also of scientific endeavour itself, questioning its worthwhileness and responsiveness to society. Dimly in the background there is some awareness of future problems of potentially disastrous proportions such as the 'genetic twilight,' as, increasingly, those with some gene abnormality now survive, by improved medical and scientific provisions, to adult life.

Fear of science is, in part, being dispelled on the one hand by means of science fiction, which provides some psychological release, and on the other by a greater openness by scientists to inform public opinion. But there is a growing desire to bring about science's integration within a more balanced society and within the wholeness of the individual.

THE THIRD PARTY IN THE DEBATE

Traditionally, the debate between science and theology (or more accurately between some scientists and some theologians) has been an affair between the two parties, with philosophers of various kinds being enlisted by both sides. In the last two decades the pressure for science to be accountable to society has grown vastly. It is held to be too important to be left solely to the scientist, and the implications of science and technology for human and social value occupy a dominant place in the assessment of scientific 'progress.' The dependence of present-day life on science calls for greater scrutiny and openness of purpose so that society may at least have an opportunity of knowing, if not judging, the relative benefits and hazards. The ball is now firmly in the scientists' court to produce justification of their activities and to deal with the difficult question of accounting for its social costs. Much of this development is welcome among the scientists and considerable progress has been made. For example, agreements have been reached and implemented on the propriety of undertaking certain classes of experiments in genetic engineering; there is a marked greater caution in the use of experimental animals and an awareness that the rapidly developing and exciting field of research in neurobiology will need to be kept under surveillance as it poses new problems.

It is in this setting that the changed public attitude to science has developed, an attitude that no longer accepts the inherent goodness of science or the priesthood of the scientist. This change was shown particularly among young people in the United Kingdom in the late 1960s when for some years there was a retreat by students from subjects such as chemistry, physics, and to a lesser extent, engineering (all thought to be causes of the 'trouble'). By contrast, medicine, biology, and sociology were more popular as desirable subjects for study. By 1980, in the United Kingdom, there had been some redress, partly stimulated by problems of another sort – the need to create prosperity to maintain a welfare state and to provide the means to share technological skills with less fortunate countries through exportation.

The scientist and the theologian have thus been joined in their debate by the socially conscientious enquirer whose concern presses for action on the benefit/hazard and ethical questions, for considerations of the quality of human life, and for assessment of the worthwhileness of technological effort. This last issue becomes continually more insistent, as scientific and technological projects become more and more costly, contending for ever larger proportions of the national wealth in competition with other urgent and socially desirable claims in, for instance, education and welfare.

It is of note that William Temple did not closely identify himself with the science–theology debates of his day, as his father had done in the previous century in the first flush of the Darwinian hypothesis. Temple's view was that of the wholeness of human experience, of which the scientific input was a small part. For him, the relationships between people and the human–divine situation were paramount. The products of science could indeed prove to be tiresome distractions. In a sense he was a man in advance of his times in foreseeing that human values and the integration of experience were of commanding importance.

In 1940 William Temple (as archbishop of York) addressed a group of young people at Doncaster, where the writer was privileged to be present. He spoke about science, education, and society in the following terms:

> Successful education means an education which enables people to live in effective relation to their environment. Science does not necessarily constitute progress. It has brought knowledge which is capable of bringing enormous benefit to mankind, but it is also capable of bringing injury to it. It is by no means clear yet that the human race is happier for the invention of flying. Ask China; ask Poland.
>
> I would give the whole of my overdrafts, and that is the only thing I have worth giving, to anyone who would 'devent' the internal combustion engine. With the exception of the train, I would do without these modern travelling facilities. But I cannot do without the railway. I would not like to travel to London by coach; but I really would like to abolish the motor-car, the flying machine, the telephone and the wireless.
>
> I suppose the time will come when the human race will find the wireless has turned out to be a blessing. Today, people are listening to the final results of a great many processes of thought, one conflicting with another, and all kinds of talks by a great many eminent persons on all kinds of specialist subjects and theories. What kind of porridge the minds of listeners are getting into I tremble to think. Science could make real progress, but did not constitute it, and the multitude of gadgets did not make the world any better or any happier.

The imaginative challenge that confronts the scientist and those who would see science's perspectives is thus much enlarged. It is no longer a matter for the laboratory innovator, or just for the philosopher or the everyday observer. It has come at this time to involve the whole community, but places special responsibility on those who guide community opinion. The theologian and the church have particular parts to fulfil, at least by insisting that questions of positive values, moral judgment, and the dignity and meaningfulness of life are central in the continuing three-cornered debate and by keeping an eye on an ever-changing scenario.

After all, it was Dean Inge who commented that he who marries the spirit of his age will be a widower in the next.

REFERENCES

Carter, G. 1980. *Higher Education*. Oxford, UK: Blackwell

Kibblewhite, E.J. 1982. In *Progress in Cosmology*, edited by A. Wolfendale. Netherlands: Reidel Press

Kolbe, H. 1877. *J. prakt. Chem.*, 15:473

Medawar, P.B., and J.H. Shelley. 1980. *Structure in Science and Art*. Amsterdam: Excerpta Medica

Monod, J. 1975. In H.A. Krebs and J.H. Shelley, *The Creative Process in Science and Medicine*. New York: Excerpta Medica/American Elsevier Publishing Co.

Peacocke, A.R. 1979. *Creation and the World of Science*. Oxford: Oxford University Press

– 1981. *Science and Theology in the 20th Century*. Stocksfield, UK: Oriel Press and Notre Dame Ind.: Notre Dame Press

Pickering, G.W. 1967. *The Challenge of Education*. London: Rationalist Press

Polanyi, Michael. 1959. *Personal Knowledge*. London: Routledge and Kegan Paul

Temple, W. 1940. Speech at Doncaster Grammar School, UK, given on 29 July

William Temple and the Future of Christian Social Ethics

ALAN M. SUGGATE

William Temple's central book on Christian social ethics, *Christianity and Social Order*, has justly been described as a key piece of writing in British Christianity. Yet it is also somewhat disappointing, in the sense that it is a writing up of what Temple had thought for twenty years and more, and rather obscures the fact that in his last years he was trying hard to be a man on the move, juggling rather untidily with several different approaches to social ethics. It is a sign of his greatness that he is most interesting at the end of his life – more interesting than *Christianity and Social Order* indicates. I want first to say a little about the contrast between the earlier and later Temple; then I shall explore the fact that he employs three different approaches to Christian social ethics: that of principles, of natural law, and of love and justice. I shall attempt some assessment of Temple, trying to do so within the terms of what was open to him in his day, (and not, I hope, with too much hindsight), and then draw out some implications for future directions in Christian social ethics.

Temple has rightly been described as a Christian philosopher, that is, one who begins his quest for a comprehension of reality from the perspective of the Christian faith. He was trained in the heyday of British Hegelianism. He conceived the philosophical task as a metaphysical one – philosophy is a 'determined effort to think clearly and comprehensively about the problems of life.'[1] For much of his life Temple was engaged in the search for a coherent account of the universe – a Christocentric metaphysic. He conceives of himself as first working from the world, through knowledge, art, morality, and religion, showing how they converge yet do not themselves meet in an all-inclusive system of truth. Then the Christian hypothesis is adopted, and he seeks to show how its central fact, the Incarnation, supplies the missing unity. In fact that perspective is there from the start, guaranteed by Temple's intense personal commitment to Christ.

Closely related to the Hegelian background is the fact that Temple imbibed the atmosphere of *Lux Mundi* and the Christian Social Union, both of which dated from 1889. While *Lux Mundi* was subtitled 'A Series of Studies in the Religion of the Incarnation,' the Christian Social Union attempted to draw out the social implications of the Incarnation. Temple, like them, saw the Logos as operative in every facet of the created world, and was disposed to look on contemporary trends in a positive manner.

Much of the inspiration for this movement came from T.H. Green, philosopher at Balliol till his early death in 1882, exponent and embodiment of what has been termed a surrogate faith, fuelled by an intense moral earnestness oriented towards social service. Green developed a highly immanentist philosophy of realization which exactly met the needs of young men from devout homes who found their faith crumbling and looked for a secular outlet for their impulses to altruism.[2] Temple never lost his hold on the transcendence of God (his *Nature, Man and God* argues strongly for it). But his inheritance gave him a disposition to believe that reality could be rationally comprehended, and to take an optimistic view of man and social institutions. He looked on sin as essentially selfishness and its antidote as sacrifice. He overrated the importance of ideas and assumed that it is possible to decide what is good in various positions and effect a synthesis, and that the interests of individual and society are capable of harmony.

These attitudes were undermined not so much by the First World War as by the economic, industrial, and political turmoil of the inter-war years both at home and abroad. By 1938 Temple was recognizing that since the Church of England Doctrine Commission began its deliberations back in 1922 there had been a tilting of the balance away from a theology of Incarnation and the search for a Christocentric metaphysic towards a theology of redemption, which was seen to be closer to the New Testament. 'We have been learning again how impotent man is to save himself, how deep and pervasive is that corruption which theologians call Original Sin.'[3] In 1939 he was calling for the digging of deeper foundations and asking what the relation is between the order of creation and the order of redemption.[4] It was one of Temple's greatest achievements to recognize the inadequacy of his attempts at a Christian philosophy and to commit himself to fresh theological tasks.

It is with this process of change in mind that we can now consider Temple's three approaches to Christian social ethics.

Temple's handling of social issues up to the 1920s is uneven. On the one hand, we find his optimistic philosophy leads him to think of Christianity as supplying ideals and the power to live up to them. His contact with a small number of working-class men in the Workers' Educational Association convinced him that the Labour movement embodied his own ideals of freedom and fellowship. He tended naïvely to declare as Christian any working-class solidarity in a strike or the miners' demand for a national pool of wages.[5] The main purpose of the Labour party, he believed, was to secure comparative equality of circumstances to citizens and to substitute the service of the community for private gain as the chief motive in industrial enterprise; and that was a Christian thing to do.[6] Apparently Labour had only to keep the ideals true and all would turn out right.[7] This thinking matches the euphoric mood of England in the period from 1918 to 1920.

On the other hand, Temple did have the wit to see that Labour solidarity rested partly on common antagonism and might collapse if it were successful against capitalism.[8] Moreover, there is a noticeable toughness in the area of international relations. As early as 1914 Temple was asking: 'Can a State obey the Christian law at all?' 'Has self-sacrifice any real meaning when applied to communities, and if so, is it in their case a virtue?'[9] Temple vigorously opposed pacifism, partly on the grounds that no direct imitation of Christ is possible for sinful men and the sinful nations of which they are inescapably members. Christ's injunctions in the Sermon on the Mount cannot be taken as absolute rules.

We find, therefore, no coherence in Temple's early thought, rather the juxtaposition of flights of idealism and sober estimates of man's predicament; intense moralism that rushes to identify the latest secular movement with Christian ideals, and hard-headed recognition of the gulf between Christ and sinful men and nations. The former is dominant in his domestic thinking, the latter on the international front.

Temple saw that he needed to tighten up. In 1920 we find him saying: 'Our duty is the very difficult one of maintaining an ideal while adopting in the most realistic manner steps which are in fact best calculated to lead to its attainment.' There is no absolute moral claim independently of all actual consequences. We must simply do the best thing for humanity in all the circumstances.[10] Because the term 'ideals' seemed to many to suggest absolute action–rules, Temple increasingly distinguished ideals from principles. The way of principles, he claimed in 1923, avoided on the one hand an ideal system conceived and implemented regardless of circumstances, and on the other a timorous acceptance of, or tinkering with, the status quo. This method is idealist in that it goes beyond the mere remedying of admitted evils and suggests positive relationships to be established; but it is realist in that it is always concerned with the application of principles to what is rather than with dreams of what might be. Then for the first time Temple co-ordinates into a set of four the social principles that he (and many other churchmen) had come to use: respect for personality; fellowship; the duty of service; the power of sacrifice.[11] This thinking was part of Temple's personal preparation for the 1924 Conference on Christian Politics, Economics and Citizenship (COPEC), which met precisely to work out Christian social principles into more specific recommendations, known as middle axioms.

This way of principles, integral to his Christian sacramental philosophy, was from now on Temple's most characteristic approach. It survived the perceptions of 1938–9 and was eventually set out in 1942 in *Christianity and Social Order*. Temple first vindicates the right of the church to interfere in the social order: the supposition of completely separate spheres of religion and politics, economics, and so on is a modern aberration. He sketches the limits of the church's competence: 'The Church must announce Christian principles and point out where the existing social order at any time is in conflict with them. It must then pass on to Christian citizens, acting in their civic capacity, the task of reshaping the existing social order in closer conformity to the principles. For at this point technical knowledge may be required and judgments of practical expediency are always required.'[12] He starts out from what he calls Primary Christian Social Principles – which are really summary Christian doctrine: God and his purpose; Man, his dignity, tragedy, and destiny. From these concepts he infers the first three of the four principles I mentioned. They express his basic understanding of personality. Personality, he writes, only becomes what it is capable of becoming through its development in the reciprocal relationships of society. Temple refuses to give primacy to either the individual or the social aspect of personality, and frequently says that our aim should be the fullest development of persons in the widest possible fellowship.[13] Armed with these principles the Christian approaches the problems of his day. Thus when Temple looks at the unemployed he writes: 'However much their physical needs may be supplied ... the gravest part of the trouble remains; they are not wanted! ... Because [a] man has no opportunity of service, he is turned in upon himself ... It has not been sufficiently appreciated that this moral isolation is the heaviest burden and the most corrosive poison associated with unemployment: not bodily hunger but social futility ... Nothing will touch the real need except to enable the man to do something which is needed by the community. For it is part of the principle of personality that we should live for one another.'[14]

These three social principles certainly do give us directions in tackling specific issues. The chief problem is that Temple has a persistent tendency to drift into moral deductions from his principles without the necessary attention to the facts of the case – not least the fact of power and the fact of sin. COPEC produced many recommendations based on principles of this kind but failed to conduct the prerequisite analysis and research. In 1926 Temple tried with others to hasten a settlement of the miners' strike. The impression is left that he was too preoccupied with the moral interpretation of events and too little alive to the alignments of power. He talks as if the principle of fellowship was bound to lead to a settlement by arbitration, but the employers had no intention of arbitration and correctly estimated there was no need for it.[15] The same problem remains in *Christianity and Social Order*. The book does speak of man's tragedy and includes a fine popular exposition of the doctrine of Original Sin. Yet this fact does not prevent an excessively moral interpretation stemming from the three social principles of respect for personality, fellowship, and service. These principles need to be set within a framework that more effectively calls attention to the facts and to sin.

It is probably the Hegelian inheritance that underlies this tendency. In formal terms Temple allowed philosophically for the consideration of circumstances. At the heart of his philosophizing is a commitment to the dialectical method as expounded by his revered master Edward Caird. As Temple put it in 1914: 'All actual thinking proceeds in circles or pendulum-swings. We approach a group of facts; they suggest a theory; in the light of the theory we get a fuller grasp of the facts; this fuller grasp suggests modifications of the theory; and so we proceed until we reach a systematic apprehension of the facts where each fits into its place.'[16] Temple has been rightly criticized by J.F. Padgett for never really carrying out the dialectical method.[17] His tendency is to short-circuit the process and produce a solution too soon. Perhaps the truth is that this formal commitment is bound to be frustrated within the Hegelian type of framework. This tendency was probably reinforced by the style of confident leadership bred into Temple by family, Rugby, and Oxford.

Let us look more closely at the transition from principles to policies. In *Christianity and Social Order* Temple moves from principles to a chapter on 'The Task before Us.' He writes that Christians cannot know what is the remedy for the problems of Britain. What they can do is call upon the government to set before itself a number of objectives and pursue them as fast as opportunity permits. He then gives six objectives (so-called middle axioms), for instance: (1) every citizen should be secure in the possession of such income as will enable him to maintain a home and bring up children in decent conditions; (2) every citizen should have a voice in the conduct of the business or industry in which he works and know that his labour furthers the well-being of the community. These objectives may seem utopian and we cannot have them all tomorrow, says Temple. But we can head in that direction in the interests of personality. This perspective brings the body of the book to an end. Temple thinks every Christian should endorse the substance of what he has been saying so far. He then gives us an appendix containing a suggested programme and makes it clear that he has moved on to disputable ground. Christians not only can but ought to contest what he says. His ideas cover greater state control of banking, land taxation, a planning authority representing all sides of industry, holidays with pay, and so forth.

It was this appendix that caused the greatest rumpus. Many critics simply ignore the disclaimers and show that they have not grasped the central thrust of the book. However, they do have a point. It will help if we consider middle axioms. The term is not a happy one and the notion has received criticism, but in my view it is still very important. Middle axioms have been a feature of ecumenical social ethics for over forty years, coming into their own at the Oxford Conference of 1937. Professor Ronald Preston has described them as 'an attempt to proceed from the basic ethical stance deriving from a theological or philosophical world-view to the realm of the empirical by seeing if there is a consensus among those with *relevant experience* of the matter under discussion (both 'experts' and 'lay' folk) as to the broad moral issues raised and the *general direction* in which social change should be worked for, without getting as far as *detailed* policies.' It is a matter first of analysing an empirical situation and then asking where the Christian understanding of life is being disregarded, and on the basis of that negative judgment forming a middle axiom. The method is a way of getting the church beyond vacuous generalities so that it can bite on events without committing itself to specific policies. Middle axioms are therefore obviously relative to time and location, and so are provisional. What weight they carry depends on how they are arrived at. Professor Preston stresses that they cannot be arrived at by clergy or by theologians alone; relevant lay experience is absolutely essential.[18]

Now the trouble with Temple's middle axioms is that they present themselves as solo efforts. True, they are presentations of views that were part of the consensus registered at COPEC and later, but here again the method was defective. There is much truth in Hensley Henson's acid remark that COPEC was the worst conceivable method of arriving at truth.[19] There was simply too little empirical analysis. Yet expert analysis is vital, and the construction of middle axioms (and any more specific positions) must be a *corporate* venture conducted by the people with the experience. This is the most responsible way of bridging the gap between lofty principle and the particularities. It avoids deductive use of the Bible or natural law and also rules out a separation of the two kingdoms. As Professor Preston writes: 'The method of the middle axiom is critical of the *status quo*, but it keeps contingent political and social judgments in their proper and necessary but secondary place. It requires Christian social action and will not sanction a private pietism, but it differentiates between God's causes and our causes. It takes the religious overtones out of politics while insisting it is a necessary area of Christian obedience.' Temple, then, was right to back the method of the middle axiom but did not back it in the best way.

The second approach Temple used to social problems was that of natural law. Temple's debt to liberal Catholicism, and especially to Charles Gore, was immense, so it is not surprising that he was influenced by the predominantly Anglo-Catholic Christendom Group to show an interest in natural law and natural order. In any case natural law is a standard feature of traditional Anglican moral theology. Through the 1920s and 1930s the Christendom Group tried to work out a 'Christian Sociology,' that is, a Christian perspective on the proper ordering of society. In 1939, when Temple was calling for the digging of deeper foundations, he expressed admiration of the papal social encyclicals *Rerum Novarum* (1891) and *Quadragesimo Anno* (1931), and of the writings of Jacques Maritain.[20]

In fact Temple had a somewhat ambivalent attitude to natural law. On the positive side he found it important because it focused attention on the right ordering of society, and on the question of ends and means. On this basis he advocated that production properly exists to meet human needs, not to maximize profits; that land owners should administer their land for the common good, not for their own profit. He believed the modern world needed an updating of the medieval doctrines of the just price and the prohibition of usury. The significance of natural law for Temple was that 'it holds together the ideal and the practical.'[21] He undoubtedly felt there was a strong resonance between the natural law tradition and his own characteristic approach through principles. A comparison of Temple with the natural law tradition in his day shows a strong affinity in substance: for instance, over the basic understanding of man as individual and communal, and over the state as having a positive role in furthering the common good, yet subject to well-defined limits. Temple's biographer tells us that he planned a personal approach to the Vatican hoping that Anglican and Roman Catholic theologians might undertake a joint study of natural law.[22]

No doubt such a gathering would have explored Temple's reservations over natural law as well. He said in 1941 that he did not think the natural law tradition faced up sufficiently to the problem of sin; and he did not believe that the Thomist doctrine of real kinds could do justice to the individuality of man. He also thought that some of Saint Thomas's ideas were more tied to the culture of the Middle Ages than practitioners of natural law cared to realize.[23] Temple was in fact far more responsive to modern thinkers than were those in the traditional mould. In good Hegelian fashion he was searching for an account of man that would combine what was valid in post-Cartesian knowledge and in the earlier tradition. It is not surprising that he warmed most to Maritain.

In my view Temple had little to learn from natural law and much to lose if he gave it too much play. What he did not see was that natural law could not deliver him from the weaknesses I have outlined in his own way of principles. This fact comes out clearly in the Malvern Conference of 1941, which was a successor to COPEC but was much more restricted in representation. It was really dominated by the Christendom Group. It ostensibly sought to be much more theological *and* to give as much practical guidance as possible for post-war reconstruction. It turned out to be very difficult to do either or both successfully. The real aim of the conference was to register agreement for practical purposes. It was therefore a major problem when arch-digger Donald Mackinnon threw up masses of earth in the faces of the assembled four hundred as he discoursed on creation, sin, and redemption (and the problematical nature of the church) in the search for foundations – precisely the task to which Temple had committed himself in 1938. Temple was caught in the trap of the main aim of the conference and found himself penning its common mind – a task which even he found very exacting.[24]

The practical quest was also unsuccessful – I would say unsuccessful precisely because the theological questions had not been faced. The Christendom Group in their passion for working out the Christian natural order neglected to take seriously the autonomous discipline of economics. Temple evidently thought they were experts, but that was simply not true.

When Temple referred at Malvern to the problem of sin he alluded to Reinhold Niebuhr. Much lies behind this remark. Just when Temple was writing his crowning philosophical work *Nature, Man and God* in the 1930s, and dedicating it to Caird, the rasping voice of Reinhold Niebuhr was troubling his peace – Temple himself so described Niebuhr and it took a lot to trouble Temple. During the thirties, in the wake of books like *Moral Man and Immoral Society*, and Niebuhr's strictures on the banal sentimentality of American

Christian liberals, Temple formulated his own views on love and justice, and gave much greater weight to the brute facts of power politics and self-interest. His position has to be fitted together out of scattered remarks, particularly on the topics of pacifism and international relations. It is a pity in my view that it was largely confined to these areas, as Temple is at his strongest here. His position is as follows.[25]

Justice is not identical with love: love *transcends* justice. For justice is a virtue relevant to the realm of claim and counter-claim, but where love is established these claims simply do not arise. This view does not mean that justice is something contrary to love which love mitigates, nor does it mean that love can leave justice behind, for love *presupposes* justice as a virtue applicable to the relations between groups. Similarly, the Gospel does not negate or leave behind the Law, but rather presupposes it. Groups function through representatives, who act as trustees for the interests of the members. It would be ridiculous for a trade union committee to prefer the interests of the employers to those of the workers. But the greater problem is that because of man's entanglement in sin neither individuals nor groups are able to fulfil the law of love. Natural communities are inevitably self-regarding; they generate a devotion that has no object outside them. This is an attribute particularly true of a nation; it is able to appeal both to the altruism and to the egoism of its citizens. The effects of this exorbitant egoism can be mitigated only if the members of the community feel they have a loyalty to a wider group. There is no effective check on national egoism and it can be demonic. Groups, therefore, especially nations, are far less amenable to the law of love than individuals. To call on nations to act by love only is likely to produce no actual result. The pacifist way of redemptive suffering, though ideally best, would in the conditions in Europe in 1939 be completely impracticable and ineffectual.

Does the law of love in its fullness, including the claim of self-sacrifice, apply to groups? Temple's answer is: 'In an ultimate sense, yes.' The idea of mutual love holds for all men and all groups, but its direct application can be irrelevant. The law of love applies indirectly to groups. Its way lies not through altruism but through reasonable claim and just reward. Justice is the true form of love at the level of groups.

Temple's position here involves two kinds of priority. The first is in terms of value. The Christian will judge all he does by the highest standards. In his thinking about corrective justice he will have as his highest aim the reformation of the prisoner as a person created as a child of God, for whom God intends eternal communion with himself. We should always aim to deal with men according to their needs rather than their deserts. But the other priority is in terms of indispensability. We have to reckon with the factor of sin and work with the material at hand. Men do not love the highest when they see it. The most fundamental requirement made of any government is not the expression of love, or even of justice, but that it should supply some reasonable measure of security against murder, robbery, and starvation. Internationally a balance of power is indispensable if nations are to co-operate in the tasks of peace. We therefore have to compromise and so order life that self-interest is harnessed and prompts what justice demands. The church's assertion of original sin, says Temple, should make it intensely realistic, and conspicuously free from utopianism. Perfect justice can never be achieved in this life. Even if men were willing to put their claims on a level there would still be the need for groups to be drawn together on a basis of common interest. And even this state should not be confused with the love of which the Gospel speaks. It is consistent but not identical with it. Moreover, perfect justice is a product of perfect love, not a stage on the way to it. Nothing short of conversion is required. We must divest ourselves of any Pelagian notion that we can build the kingdom of God. The goal of Christian hope is not any kind of social or political achievement, and its realization lies beyond the present conditions of mortality.

Temple undoubtedly owed much to Reinhold Niebuhr for this basic position, but it is only fair to say that much of it can be found in different words in his earlier writing. Now there are two points, very closely related, at which Temple diverges from Niebuhr. The first is that whereas Temple construes love primarily in terms of mutuality, and either subordinates or merges sacrifice with it, Niebuhr construes *agape* as essentially sacrificial, and tends to see mutual love as having the root of selfishness in it. It is true that he does speak of love in terms of mutuality: the law of man's nature is love, the harmonious relation of life to life in obedience to the divine centre and source of man's life.[26] But Niebuhr's preoccupation is with the confrontation of that law of love with this fallen world. Heedless uncalculating love must entail self-sacrifice in this life. 'The perfect disinterestedness of the divine love can have a counterpart in history only in a life which ends tragically, because it refuses to participate in the claims and counter-claims of historical existence ... A love which seeketh not its own is not able to maintain itself in historical society.'[27]

Such a conception of love means, secondly, that it is impossible to construct a social ethic out of the ideal of love in its pure form. Like Temple, Niebuhr neither identifies nor divorces love and justice. Love transcends justice, fulfilling and redeeming it; love requires and presupposes justice. But Niebuhr plainly insists that love also *negates* justice. Whatever our achievements in the realm of justice, they always stand under the judgment of love; for the laws of justice take sinful self-interest for granted and are therefore always in danger of throwing the aura of moral sanctity upon it. They must stand under the criticism of the law of love. 'There is no justice, even in a sinful world, which can be regarded as finally normative. The higher possibilities of love, which at once is the fulfilment and the negation of justice, always hover over every system of justice.'[28]

Now I suppose Temple does almost reach this idea of love negating justice. For instance, in November 1939 he said in a letter to a friend on the pacifist issue: 'When I say that in the circumstances killing is right, I am not denying that it is sinful ... We have to do the best we can being what we are, in the circumstances where we are – and then God be merciful to us sinners!'[29] But he never says explicitly that love negates justice. Though he enters anti-utopian caveats, the impression received is that we are on a sloping road at the bottom of which is complete selfishness and at the top complete mutuality, and that we can choose between ascending through different levels of justice towards love, or descending through vindictiveness to chaos. Here we see the Platonic cast of mind leavened by a Hegelian propensity for synthesis and optimism. It is also true that traditional Anglican moral theology tended to see love and justice as two tiers in the same building, not dialectically. Temple's heritage made it very difficult for him to respond adequately to Niebuhr. He wanted to give greater weight to a doctrine of redemption. Niebuhr offered him a dialectical way of relating love and justice that was rooted in the Cross. Temple does not really rise to this conception, and the prevailing impression is that he was in the end trying to adjust his existing thought and accommodate Niebuhr into it. Mystification as well as mirth is perhaps packed into that limerick ascribed to Temple:

> At Swanwick, when Niebuhr had quit it,
> A young fellow said, 'Now I've hit it –
> Since I cannot do right,
> I must find out tonight
> The right sin to commit, and commit it!'[30]

It should now be clear why I said Temple was trying to be a man on the move, juggling untidily with different approaches. References to natural law and love and justice are of course present in *Christianity and Social Order*, but the way of principles dominates, and no attempt is made to explore the other ways and reach a coherent position; it is more a case of juxtaposition.

What can we learn from this investigation? I want to float two sets of tentative suggestions. The first relates to the basic conceptual tools of Christian social ethics, the second to the method of corporate reflection. If Temple's chief weaknesses are inadequate attention to the empirical and a defective grasp of the problems of sin and power, I believe that, without being uncritical, we can work along Niebuhr's lines in constructing a more adequate theory of Christian social ethics.[31] Niebuhr has certainly not been superseded – not even by liberation theology. I suggest that we should place at the centre a view of love and justice that springs from a basic understanding of the Christian faith, symbolized centrally by the doctrines of God, Creation, Incarnation, and Redemption. We must allow adequately for the eschatological fact that the created order on the one hand is redeemed and yet on the other is still subject to sin and death, and thus we must distinguish clearly between an ultimate state where love and justice coalesce in perfect mutuality and our present state where perfect love is seen in the sacrifice of the Cross and love and justice are seen in constant tension. Niebuhr's evangelical-reformed background makes him hard for many to digest, but he is still compulsory reading. Presidents and prime ministers ought not to be let loose until they have done a course on Niebuhr.

Further, we should have as clear and coherent a view of man as possible. I suggest that three aspects should bear primary stress. (1) Man is a creature standing before the transcendent God, capable of response to him, redeemed by him, and intended for eternal communion with him. By God's grace man has indeterminate possibilities of envisaging goals and shaping nature and history. This view warns us against imprisoning man in any immanentist philosophy, and against cynical and pessimistic forms of realism. This side is covered very well by recent theologies of hope. (2) Man is finite, embedded by God in nature and history. This awareness forces on our attention the fact that man's vision will always be limited. It alerts us to acknowledge a high degree of cultural relativity. We have to give full weight to the facts of the case and to the autonomous perspectives of empirical disciplines like the sciences, economics, and the social sciences. (3) Man is a sinner, always in need of the mercy of God. Few men have been so sensitive as Niebuhr to the capacity of men for egoistic pretensions, using reason as an instrument of their interests. We must also reckon with the individual and communal aspects of man – Temple's social principles are perfectly valid, though inadequate if taken on their own.

Secondly, I want to pick up the references to corporate reflection on social problems. Temple himself did see the importance of the corporate approach when in the 1930s he took the initiative to convene a small interdisciplinary group to make a thorough study of the psychological, social, and moral effects of long-term unemployment. The investigators put a strong accent on personal interview. This study resulted in a highly acclaimed volume *Men without Work*.[32] Temple rightly called it both scientific and human, and it accelerated the growth of occupational centres and gave Temple a solid base on which to argue for family allowances. One of the participants, W.F. Oakeshott, remarked later: 'I am inclined to think that the enquiry did represent what would have been a new departure in method for him – though the immediate threat of war prevented this from being at all widely employed – viz. the creation of small groups of men to think out the relations of this, that, or the other social problem.'[33] Since Temple's day the method of corporate exploration has been very widely employed by the World Council of Churches and in the several churches, especially at the level of boards of social responsibility.

I believe there is a great need for this method of corporate reflection to be extended in some appropriate way to the local level too, encouraging men and women to come together to consider daily problems and opportunities. I know from experience that there are many Christians who find it difficult to see how their faith can be a resource to them in daily living, especially if they belong to large institutions. Some find they are caught up in rapid change and face an uncertain future; others are conscious of a tension between a Gospel of love and the rules within which they are expected to work as a foreman or a manager. All too often they find that few of their colleagues are Christians, and that their parish is concerned only with the domestic aspects of Christian living. My experience has taught me the value of small groups that can sustain you in the search for a Christian response to opportunities and problems of daily life – a value that reaches far beyond the intellectual. Here I am in the debt of Bishop Ian Ramsey, noted for his interest in disclosures and stimulator of corporate interdisciplinary enquiries into many issues. In 1969 he appointed a theological consultant in industrial and social affairs, Miss Margaret Kane, and it was she who gathered together groups of people struggling to make connections between faith and life.

If such groups are to perform their task well they need to be firstly ecumenical, to give scope for a variety of Christian perspectives; and also to consist of clergy *and predominantly* laity. As far as possible they should draw in non-Christians, who can help to explode esoteric religious language and keep sights on the particular situation, besides bringing their own experience and perspective. Groups of this kind also need people of varying expertise. A vital person here is the theologian, who will contribute from his or her knowledge of the Christian tradition. He will need to develop the skill of deciding what in the tradition really does illuminate the situation under discussion. It is usually best to start out from a quite specific situation concerning members and to let them tell their story so that it is understood in some detail. Then comes the time to see how the experts present can throw light on the story. In this way each member is enabled to make his own connection between faith and life, and to decide what are the next practical steps to be taken.

In making these suggestions I am still left with many puzzling questions, some of which James Gustafson has recently indicated.[34] How are we to deal with polarities like nature and grace, creation and redemption, continuity and change? What is to be our fundamental understanding of God? How are we to establish hermeneutical principles for using the Bible in theological ethics? How can we do justice to the insights of historicism and existentialism and to the need for general moral principles – perhaps a revised natural law? How can we achieve a coherent and adequate theological ethics, responsive to current problems and responsible to moral values and principles, grounded in the faith and life of the Christian community and in our common humanity? How do the insights and models of the social sciences play their part in decisions in social morality? How do we best encourage ordinary Christians to find in their faith a resource for Christian living? If we are to make progress in a very complex field it is clearly essential for professional theologians to meet academics of other disciplines. But theology must not be left to the academics. All of us who are concerned with the relationship of Christian faith to life are involved in the process of theology and we all have much to learn from each other.[35]

NOTES

1 *Mens Creatrix* (London: Macmillan 1917), 7
2 See M. Richter's excellent study of Green, *The Politics of Conscience* (London: Weidenfeld and Nicolson 1964).
3 *Doctrine in the Church of England* (London: SPCK 1938), 16–17
4 'Theology To-day,' *Theology* (Oct. 1939); reprinted in *Thoughts in War-time* (London: Macmillan 1940), 93–107
5 *The Challenge*, 11 Jan. 1918; *The Pilgrim*, July 1921, 365–6
6 *Chronicle of Convocation*, 1918, 349–53
7 *The Daily Herald*, 26 March 1921
8 *Essays in Christian Politics* (London: Longmans Green 1927), 5
9 *Christianity and War* (London: Oxford University Press 1914), 10–13
10 *The Contemporary Review*, July 1920, 65–70
11 *The Pilgrim*, January 1923, 218–25
12 *Christianity and Social Order* (Harmondsworth, UK: Penguin 1942), 35
13 E.g., *The Church Looks Forward* (London: Macmillan 1944), 118
14 *Christianity and Social Order*, 12
15 See Temple's letter in *The Times*, 21 Aug. 1926; on the employers, see G.W. McDonald, 'The Role of British Industry in 1926,' in M. Morris, *The General Strike* (Harmondsworth, UK: Penguin 1976)
16 *Studies in the Spirit and Truth of Christianity* (London: Macmillan 1914), 42–3
17 J.F. Padgett, *The Christian Philosophy of William Temple* (The Hague: Nijhoff 1974), 235
18 R.H. Preston in his introduction to the reissue of *Christianity and Social Order* (London: Shepheard-Walwyn and SPCK 1976), 8; and in 'Middle Axioms in Christian Social Ethics,' in *Crucible*, January 1971, 9–15
19 Henson Papers, quoted in E.R. Norman, *Church and Society in England 1770–1970* (Oxford: Clarendon Press 1976), 309
20 *Thoughts in War-time*, 105
21 *Christianity and Social Order* (Harmondsworth, UK: Penguin 1942), 57–61; *The Hope of a New World* (London: SCM Press 1940), 67; *Religious Experience* (London: Clarke 1958), 231

22 F.A. Iremonger, *William Temple, Archbishop of Canterbury* (London: Oxford University Press 1948), 423–4

23 *Malvern, 1941: The Life of the Church and the Order of Society; Being the Proceedings of the Archbishop of York's Conference* (London: Longmans Green 1941), 14–15; *Religious Experience*, 231–6

24 *Malvern, 1941*, especially D.M. Mackinnon's article, 81–116

25 The chief sources are: Temple, 'Christian Faith and the Common Life,' in work of the same title, vol. 4 of series from the the Oxford Conference, 1937, on Church, Community and State (London: Allen and Unwin 1938); *Christianity in Thought and Practice*, esp. ch. 3 (London: SCM Press 1936); *The Hope of a New World* (London: SCM Press 1940); *Citizen and Churchman* (London: Eyre and Spottiswoode 1941); *Thoughts in War-time*; *The Ethics of Penal Action* (London: National Association of Penal Officers 1934); together with articles in *The Fortnightly* and *The Christian Century*, and letters and reports in *The Times*.

26 R. Niebuhr, *The Nature and Destiny of Man* (London: Nisbet 1941), 1:17

27 Ibid., 2 (1943): 75

28 Ibid., 1:302; 'Christian Faith and the Common Life,' 72 (see n. 25)

29 Quoted in Iremonger, *William Temple*, 542–3

30 Quoted in H.S. Leiper, 'William Temple – Man and Friend,' in *Christendom* (US) 10 (winter 1945), 11

31 See G. Garland, *The Thoughts of Reinhold Niebuhr* (New York: Oxford University Press 1960)

32 *Men without Work* [with an introduction by Temple] (London: Pilgrim Trust 1938)

33 Quoted by Iremonger, *William Temple*, 442–3

34 J. Gustafson, *Protestant and Roman Catholic Ethics* (Chicago: University of Chicago Press 1978; London: SCM Press 1979), 31–3, 60–2, and ch. 5

35 For amplification of points made in this talk, see my 'William Temple's Christian Social Ethics: A Study in Method' (PHD dissertation, Durham, UK 1980). Part of this talk has already appeared as part of the article 'William Temple and the Challenge of Reinhold Neibuhr' in *Theology* (Nov. 1981): 413–20.

Part and Apart: Issues in Humankind's Relationship to the Natural World[1]

ROBERT W. KATES

For most of human history, science and religion, broadly defined, pursued a common task as part of 'the experiment of life,' probing and defining humankind's relationships to the natural world. In the last century that intermingling of effort has diminished and for reasons I do not fully understand, environmental issues are debated by those affected, by government, by environmental activists, and by scientists but seldom by the religious community. In this paper I describe the varied and conflicting images we hold of our relationship to nature, and the roles played by science and religion in their conceptualization. I then speculate on some reasons for the demise of interest in these matters and conclude with an extensive case study from my research on technological hazards, suggesting the kinds of issues requiring a collaborative effort by science and religion.

THE NATURE OF NATURE

The intermingling of science and religion in 'the experiment of life' is clearly revealed in the involved history of humankind's relationship to the natural world. Together we came to consciousness of a natural world separate from ourselves. Together we probed its mysteries and sought to exploit its humanly defined resources. Together we sought to map the complex of systems we call nature and by ordering the parts we sought to grasp the whole. We shared in each of these endeavours while possessing a common flaw, albeit one that would appeal to the dialectician in William Temple. We are both part of and apart from the natural world.

Part and apart

It is the legacy of human consciousness to be forever trapped in such dualities as subject/object, observer/observed, or participant/observer – to be as one with our world, intimate and a part of our world by virtue of our being and separate and apart from it by virtue of our consciousness and reflection. Thus ecologists may tell us of a seamless web of life and poets may sing of it. Yet our own words, scientific and poetic, speak of fundamental contradiction. In our very act of affirming its unity we conceive of and indeed bemoan the separation of humankind from the natural world and by implication the separation of people from their environment.

The dilemma is not effectively bridged either by scientific model or poetic metre. We are apart from nature, forever so, because man and nature are different words and whether we insert hyphens, arrows, or feedback loops, the differentiation still stands. Yet in some special way we know that both the ecologists and poets are right. We are trapped in the reductionism of words and thought at the very moment that we aspire for organic unity.

In the beginning, perhaps, this was not our state. There was a time when nature was just a given. But three other major conceptual images of nature emerged to dominate the uneasy human relationship with the natural world. Nature is viewed variously: as given, as mystery, as dominion, and as system. Each is rooted in our mythic human origin and each persists today.

Nature as given

It is probable that in the beginning, just as today, the most common conceptualization of nature was as given. Nature was synonymous with environment, literally that which surrounds. The need to differentiate nature from artificial environment had to await the creation of a built environment. To the external observer, the human relationship to nature was surely one of dependency – nature was and is the basic life-support system for all species. How conscious our ancestors were of that relationship is of course unknown, but not totally so.

To ask what might have been the conceptualization of nature among the earliest of our human ancestors is not to engage in what Evans-Pritchard (1965) mockingly calls the 'if I were a horse' fallacy – acts of imaginative fancy and projection beyond the realm of credible verification. We have after all some knowledge and some informed speculation about the material basis for life of our ancestors and the observations of current or recent human groups whose livelihood of hunting and gathering surely bears some resemblance to that of our earliest ancestors. We also have a small but growing understanding of the perception of environing objects by infants, which involves a process of differentiation of self and surrounds perhaps not unlike the historic phylogenetic process. Finally, we can draw upon our own experience of humanness to project backwards and still be somewhat more hopeful of results than in projecting onto horses.

From such data, observations, and experience we might conclude that in the beginning the dominant image of nature was as a given – it was there. But consciousness, and thus differentiation from self and species, must quickly evolve. Such consciousness focused on specific natural elements, not on nature as a whole. These would likely be elements that sustained or threatened life but were not so all-enveloping as to be taken for granted (for instance, food as opposed to air). Lucretius seems to capture the distinction as follows.

> When overtaken by night they [the human race] laid their naked bodies on the ground like bristly boars, rolling themselves in leaves and foliage. Nor did they go wailing through the fields to seek the day and the sunlight, fearing the shadows of the night; but quietly and buried in sleep they waited until the sun with red torch should bring light to the sky. For since from infancy they had been used to seeing darkness and sunlight born in

turn, it could not occur to them ever to wonder or to doubt whether eternal night might not hold the earth forever and the light of the sun be withdrawn. But they were troubled rather about the herds of beasts, which often made rest dangerous to these unhappy beings. Driven from home they fled their rocky shelters at the coming of a foaming boar or mighty lion, and at dead of night yielded in terror their leaf-strewn beds to their savage guests. (*De rerum natura* 5, 925–1010; quoted in Lovejoy 1935, 228 – see References at end of this paper)

In terms of direct paleontological evidence there is little for conceptualization of nature until the upper Paleolithic (ca. 40,000 BC) whose 'cave engravings and paintings [and] the burials with their widely varying peculiarities ... [offer] evidence for a complicated and rich ideological world' (Blanc 1961, 119).

How much earlier, preceding the pictorial records of the cave with its magical evocation of the hunt, differentiation did take place is not clear. Some speculations have a quality of immediacy provided by an imaginative (perhaps unbelievable) projection backward of current philosophical concerns.

In the absence of documents (apart from the tools of the 'pebble culture'), it is necessary to try to recapture the psychological state of these first mutants, suddenly separated from the other primates, taking their first hesitating steps in an unknown world ...

The animals surrounding him [mankind] came and went, indefatigably repeating the same actions: hunting, gathering, searching for water, doubling or fleeing to defend themselves against innumerable enemies; for them, periods of rest and activity succeed each other in an unchanging rhythm fixed by the needs for food or sleep, reproduction or protection. Man detaches himself from his surroundings; he feels alone, abandoned, ignorant of everything except that he knows nothing, no longer forced to obey the laws of the clan, from which he feels irremediably cut off. His first feeling thus was existential anxiety, which may even have taken him to the limits of despair. Without previous experience, his consciousness was necessarily rudimentary and rough, yet it was an authentic human consciousness. (Bergounioux 1961, 110–11)

Perhaps it is easier to speculate on infant conceptualizations as an analogy with human evolution. Again the point at which differentiation between self and surrounds takes place is never known, nor how those surrounds are conceptualized. Instinctual responses to outside stimuli take place early, even before birth, and taste preferences are experimentally in evidence shortly afterwards (Jackson and Jackson 1978). By eight to ten weeks after birth there is recognition by infants of a human face (or facsimile thereof) followed by a growing ability to differentiate their own body from others. At approximately five months, the child has a growing ability to differentiate his or her own body from the body of the mother or primary care givers (Kaplan 1978). Active grasping for objects within range of the hands begins early and is mastered by the fifth or sixth month of life. By the tenth month exploration on hands and knees usually takes place, and the full-blown exploration of the world of subsistence work in the course of play or chores may begin as early as year two (Fraiberg 1968). The dominant concerns of the child are with needs, but more than what is needed is explored, touched, and briefly questioned. This explorative curiosity surely applies to adult persons and must have characterized early questioning of nature beyond the elements of need. Thus starlight might be observed and speculated upon by early humans irrespective of actual or supposed functional need.

The acceptance of nature as a given does not imply a lack of knowledge. To be born or to evolve into a sustaining environment, and to then fix on dominant objects or elements based on need and occasional curiosity, easily leads to careful observation and growing knowledge of the needed elements. Elaborate ethnoscientific taxonomies can be elicited in non-literate societies relating to needed natural elements (Conklin 1980). In fact such common knowledge (as opposed to specialist knowledge) actually diminishes with increased technological sophistication (Johnson 1977). The ethnoscientific data also strengthen the view of nature as a given. In a series of seminars held with groups of San-speaking hunters (Bushmen) of Namibia and Botswana, Blurton-Jones and Konner (1976) found their informants had an elaborate ethology of animal behaviour, very strong on observations and description, but very weak on explanation – a position consistent with a 'given' view of nature. Finally, the ethnoscientific data also suggest that the common knowledge is limited to the observable (Page and Richards 1977). Other conceptualizations of nature are needed for that which cannot be seen, either in time past or below or beyond the range of vision.

Nature as mystery

The consciousness of what in nature is needed for life is probably inextricably bound to what is needed for wonder. It is easy to speculate on the origins of wonder: in the fluctuation of supplies of food or water, in human disease and the experience of death, in curious eyes examining star movement or tree growth. Nature as provocateur of wonder is the second major conceptualization – a nature of wonder, of mystery, of sacred object and symbol, and of mystical, singular wholeness.

All of the major 'scientific' (as opposed to theological) theories of early religion appear to be functional and utilitarian. Religion in Evans-Pritchard's summary (1965) arises out of the need to interpret natural phenomena, to discharge emotion, or to maintain social cohesion and community. And ecologically oriented anthropologists would add to the list the maintenance of ecological balance (Rappaport 1968; Harris 1979). Excluded from this rational tradition is the approach of simply taking the religious expression at its word, or of accepting 'supernatural theory.' Setting aside for the moment (but not forgetting) this literal interpretation, we find the utilitarian explanations in conflict as to the role of nature in the world of spirit.

The utilitarian theories provide for a trichotomy of explanation. For example, animistic belief (existence of a soul in natural objects) is explained as either (1) a rationally intended but mistaken proto-scientific explanation for natural process, or (2) a convenient symbol or totem for various psychological or social needs, or (3) a way of regulating human behaviour for the wise use of the natural objects. Similarly, religious imperatives such as the dietary abominations of Leviticus are interpreted as either (1) reflecting the proto-scientific taxonomies of a biblical ethno-science (Douglas 1966), or (2) as a way of differentiating Jews from their neighbours (Douglas 1972; Soler 1979), or (3) as encouraging a utilitarian benefit by outlawing the pig for public health (Maimonides 1881; Ptai 1978) or ecological reasons (Harris 1966).

This lively debate about the origin and function of religious practice may be trans-scientific, as Evans-Pritchard claims, beyond the realm of scientific evidence. But regardless of its origin and function, the phenomenon of nature as wonder and mystery evolves in historic (literate) time in two significant directions. A concept of holistic nature rather than an assembly of natural objects or phenomena emerges. And a Western society majority opts for deciphering the mystery of nature, while most of the world is still immersed in it.

Somewhere in time nature becomes Nature, individual natural elements become a collection of elements and become personified (nature herself, Mother Nature) or systematized (laws of nature). For Williams (1976) 'Nature is perhaps the most complex word in the language [with] three areas of meaning: (i) the essential quality and character of something; (ii) the inherent force which directs either the world or human beings or both; (iii) the material world itself, taken as including or not including human beings.' For both the second and third meanings, nature must become whole, and the earliest sense of wholeness was most probably linked to pantheistic religious expressions, of unitary god(s) who were in, a part of, or encompassing the world.

The great expressions of such beliefs are in the Orient – Hinduism, Buddhism, and Taoism – but there is a Western minority tradition from the nineteenth century onwards in which some British Romantic poets and American transcendentalists find comfort. The culmination of this trend is in the sacralization of pristine nature as wilderness in the late nineteenth century (Graber 1976).

The unity of person and nature in the spirit or the solitude of wilderness does not in itself augur an unchanged natural world. As Tuan (1968) has described, the most humanized of landscapes are found amidst the strongest of oriental unitary and sacred conceptions of nature. But it is in the West that the concept of nature as mystery, although it may be ever-present, is but a muted counterpoint to the prevailing vision of a subservient nature.

Nature as dominion

The theme of nature as human dominion surely precedes the Judaeo-Christian tradition, but nowhere is it given such clarity or such eloquence as in the cosmological myth of the Western world.

> And God said, Let us make man in our image and after our likeness: and let them have dominion over the fish of the sea, and over the fowl of the air, and over the cattle, and over all the earth, and over every creeping thing that creepeth upon the earth. So God created man in his own image, in the image of God created he him; male and female created he them. And God blessed them, and God said unto them, Be fruitful, and multiply, and replenish the earth, and subdue it: and have dominion over the fish of the sea, and over the fowl of the air, and over every living thing that moveth upon the earth. (Genesis 1:26–8; King James Version)

In a powerful essay, the medieval historian Lynn White Jr (1967) found in this concept of dominion the historical roots of our ecologic crisis. The response to his essay, viewed a decade or more later, seems remarkable in the heat it evoked. In his own words: 'I was denounced, not only in print but also on scraps of brown paper thrust anonymously into envelopes, as a junior Anti-Christ, probably in the Kremlin's pay, bent on destroying the true faith' (White 1973, 60).

A flood of literature followed (Schaeffer 1970; Black 1970; Santmire 1970; Barbour 1972, 1973; Passmore 1974) mostly questioning or modifying the White thesis. Two major criticisms emerged. First, the biblical grant of dominion consecrated but did not originate human domination. The manipulation of nature by humankind was well under way about three thousand years ago when the Genesis myth was first set to paper (Passmore 1974). And secondly, White ignored the alternative or complementary role of stewardship posed in the Judaeo-Christian tradition (Black 1970). Passmore queries how really deep the role of stewardship is in that tradition, but does not question the existence of the notion of humankind being God's deputy on earth entrusted with the well-being of the dominion, or the companion Judaeo-Christian idea of co-operation with nature in its improvement. But stewardship and co-operation are clearly minority traditions, and regardless of whether the grant of dominion came before or after the fact of its exercise, the acceptance of human dominion over nature is clear and powerful in Western thought, reinforced by the Baconian–Cartesian revolution (Passmore 1974; Glacken 1967; and Leiss 1972).

For Bacon, 'Man, by the fall, lost at once his state of innocence and his empire over creation, both of which can be partially recovered even in this life, the first by religion and faith, the second by the arts and sciences' (quoted in Glacken 1967, 472). Arts and sciences, in Descartes's words, provide the means of 'ascertaining the force and the action of fire, water, the air, the heavenly bodies, and the skies, of all the physical things that surround us ... and thus make ourselves the lords and masters of nature' (quoted in Glacken 1967, 477).

For the masters and possessors of nature, the task was to decipher its mystery. In that effort a complex, expanding systemic view of nature would emerge. In the Baconian–Cartesian equation of power and knowledge it would strengthen the sense of dominion, but in the wonder and awe of scientific pursuit a new sense of mystery would emerge.

Nature as system

It is widely accepted by students of humanity that most if not all people have systems of belief and thought akin to science. These systems, often differing fundamentally from modern Western science, are collectively labelled ethnoscience, presumably to link them to the ethnographies of which they are a part. Such sciences consist of combinations of theory, method, and fact that purport to observe, classify, and explain the phenomena of nature in an orderly or systematic way. If ethnoscience is almost universal today, surely some prehistoric people would have placed their considerable knowledge of the natural world into a systematic framework somehow linking themselves to the objects or organisms that surround them in a series of functional purposes and relationships.

Central to that linkage may have been God or gods, for as Glacken (1967) notes: 'In ancient or modern times alike, theology and geography have often been closely related studies because they meet at crucial points of human curiosity.' Thus, he goes on,

> The conception of the earth as an orderly harmonious whole, fashioned either for man himself or, less anthropocentrically, for the sake of all life, must be a very ancient one; probably we must seek its ultimate origin in earlier beliefs in the direct personal intervention of the gods in human affairs or in the personification of natural processes in the naming of gods of the crops, and in the old myth of the earth-mother. (p. 36)

And of the Hellenistic age, Glacken wrote: 'The idea that there is a unity and a harmony in nature is probably the most important idea, in its effect on geographical thought, that we have received from the Greeks, even if among them there was no unanimity regarding the nature of this unity and harmony.'

Greek ethnoscience was to range widely over natural phenomena, from a credible (by modern Western science standards) exposition of the hydrologic cycle by Anaxagoras, through the elements of nature of Empedocles (air, earth, fire, and water), to the atomic theories of Leucippus and Democritus.

Further elaboration of nature as a system accompanied the birth of modern science, particularly in astronomy but much less so in natural history and earth science, and it was not until

> a group of writers, most of them living in the seventeenth century, and none in the front rank with Newton, Descartes, or Galileo, became interested in natural history, physico-theology, and scientific research, that these

> inquiries were identified with further discoveries of the wisdom of the Creator in his individual productions of nature and in the interrelationships he had established among them. Such studies and interpretations of living nature as a whole became the basis for modern ideas of the unity of nature advanced by such men as Count Buffon in the eighteenth century and Darwin in the nineteenth century. Darwinism in turn led to the concept of balance and harmony in nature, the web of life, and then to the recent concept of an ecosystem. (Glacken 1976, 379)

Physico-theology, the study of God in the works of nature, is best exemplified in the work of John Ray (1759) whose *The Wisdom of God Manifested in the Works of Creation* was first published in 1697. The argument is that natural phenomena bespeak a sense of order and purpose that can be properly explained only as the work of a divine artisan. In a famous, often repeated analogy, Ray describes the inferences that must result if we were to discover a clock in what was thought to be a totally uninhabited area. An examination of the clock reveals manufacturing skill and an orderly relationship of parts which work together for a common purpose. We would be forced to conclude that the clock was indeed an artefact, consciously assembled by an intelligent being whose existence must be postulated to account for the clock's presence. Ray then argues that many features of the natural environment have similar characteristics and thus are evidence for the existence of God.

Physico-theology developed gradually in the early Christian church, partially in response to a similar tradition in Greek thought of which Plato's *Timaeus* is the best representative. Ray's example is typical of the Christian physico-theologists in its reliance upon teleology, or the doctrine that there is an ultimate purpose or goal to nature. However, it is not necessary to accept teleology in order to believe that nature is a divine manifestation, and vice versa.

The physico-theologists were not original scientists – at best they were amateurs – but they were zealous synthesizers. Ray's friend William Derham (1798) worked out elementary concepts of the food chain, of the interdependence of all forms of organic life, of the distribution of land forms, and of the role played by physical agents such as streams and winds, and the climatic effects of earth position and axis tilt. Ray anticipated and influenced Linnaeus whose extraordinary acts of ordering, classifying, and naming made real the orderly world of the physico-theologian and laid the basis for extending the web of nature into time as well as space (Eiseley 1958). This would require discarding 'the great chain of being' or 'the scale of nature,' on which all life had been placed and created simultaneously. Time would have to be extended past biblical time. And biological mechanisms for variation and selection would have to be proposed – all of these developments taking place in Darwin's century (Eiseley 1958).

Towards the end of that century, the formal concept that living organisms and their environment comprise an interacting system with well-defined properties emerges in the works of Möbius, Forbes, and others. 'Ecology' was proposed by Haeckel in 1869, and 'ecosystem' by Tansley (1935). And finally the natural system would serve as a model for the abstraction of all systems (von Bertalanffy 1950; Miller 1978).

Image and praxis

In the flow of human thought, these four themes resemble the braided streams of old deltas rather than the incised waters of young uplands. Now diverging, now converging they carry an enormous variety of human concerns.

These flows of thought rise far upstream in the earliest human beginnings. And despite their somewhat chronological presentation, the first rivulets probably arise almost simultaneously. Yet as ancient as their origins are, they are fresh and relevant to the environmental issues and theories of the day.

In their conscious acting upon nature, people choose one of four major alternatives: to preserve, conserve, exploit, or re-create elements of the natural world. Each alternative, in turn, is strongly influenced by one or more conceptualizations of nature and of our divided relationship with the natural world.

By aboriginal fire, by land clearing and draining, by diverting the water courses in their arid margins, by working and reworking the soils, by paving the land, by domesticating plants and animals, and by diffusing or destroying habitats and species, a continuous act of creation unfolds. Much of it is done as it has always been done, with nature as given, in small incremental activities of necessity or occasional curiosity, whose cumulative effect remakes the soils and landscapes of large areas of the world. Such small interventions may retain considerable communion with nature – even today they are consciously seen as carrying out God's design or a secular 'design with nature' (McHarg 1969). Nevertheless today great fear is expressed that such cumulative actions threaten the lives and livelihoods of many in the habitats of the desert margins, mountain foothills, and tropical forests.

The most conscious expressions of dominion are in the form of large-scale works of plan and design. These are expressions of mastery and power, or at the very least the improvement upon and continuance of divine intention by scientific logic or technological possibility. They are not unthinking interventions in the natural world. There are few large-scale endeavours undertaken today in which sophisticated efforts are not made to understand and anticipate the impacts of the new creations. For the new lakes behind tropical dams, the diverted river courses of the Arctic, or the heat islands of urban electric generating stations, efforts are made to trace the linkages to pre-existing natural and human systems and to project the complex interrelationships that will develop. In such undertakings, however, nature is surely objectified and our apartness from it is strongly expressed.

Re-creation overlaps with use. The changes that occur with use are smaller, slower, and perhaps reversible. Exploitation, conservation, and preservation form a continuum of use in which the rate of extinction or pollution determines how we classify the behaviour. In exploitation the change may be irreversible – ores are lost forever, extracted as quickly as energy or economy permits; air and water serve as natural chimneys and sewers destroying the life dependent on their pristine conditions. In conservation the rate of exploitation is adjusted to some optimal level thought to sustain the yield of flow resources or to leave some legacy of stock resources to the future. To leave a full legacy, if only as the option for future exploitation, preservation sets aside wilderness areas, creates natural reserves, and rescues endangered species.

Underlying these choices is an overlapping of concepts as dominion leads to exploitation, stewardship or system to conservation, system and mystery to preservation. In both exploitation and conservation we inevitably stand apart from nature. Only in preservation can we seek to reclaim our organic place in nature and both images of system and mystery encourage us to do so. Almost all of us experience dominion and mystery, apart and part, simultaneously. In dominion we affirm our separateness, in mystery we seek unity, and in system we hope somehow to find a promised bridge.

Science and religion diverge

This then is my recounting of the emergence of four distinct views of the human relationship to nature. Throughout my tale, religion and science are intimately involved, often in one and the same person. But in the post-Darwinian age the weaving of science and religion unravels.

Thus today, although religiously based charitable and social action groups abound, there are few that are dedicated to environmental action. For example, in a recent listing (National Wildlife Federation 1978) of 378 Canadian and U.S. citizen groups and international agencies concerned with natural resource use and management only one appears to have a religious base: the St Hubert Society of America, dedicated to the principles and traditions of good sportmanship and hunting.

Except for a flurry of outrage at Lynn White's indictment of the Judaeo-Christian tradition as the root of the ecological crisis (1967), theology says little in the way of advocacy for nature. When the church in my country does take a stand on issues related to the natural world, for example in the nuclear power debate (National Council of the Churches of Christ 1974, 1975), the advocacy relates primarily to issues of war, economic injustice, and human health.[2] This emphasis is in sharp contrast with the stand of scientists such as Rachel Carson, Barry Commoner, and Paul Ehrlich who emerged as powerful advocates and strong allies of the secular nature spiritualists and the emerging environmental movement.

Why in this century after William Temple's birth the intermingled roles of science and religion in defining the human relationship to the natural world should be abandoned is beyond my knowledge but may emerge in the proceedings of this conference. Indeed, not everyone will acknowledge that it has been abandoned. Our conference chairman, for example, tells me that there is a continuing involvement in environmental issues within the church in Canada and Dr Peacocke has given me examples of interest in England and Scotland. Nonetheless there has been a diminution of interest overall and I can at least speculate upon some causes.

By Darwin's time the 'gifted amateur' of both science and religion gives way to the age of specialization in each field. Churchmen no longer write the natural history of places such as Selborne (White 1897). Some of the church goes into opposition with science when evolutionary theory appears to undermine the literal statements of the sacred myths, and other segments go into isolation; there are separate domains for science and religion. Finally, a powerful secular spiritualism based on nature develops its own mystique at the same time as the worldly concerns of the organized religions seem to focus on the relationship between people and social classes, ethnic groups, and nations.

But surely the task is not completed, and an enduring order, natural or otherwise, has not been achieved. Our conflicting relationship with the natural world persists, heightened by our appreciation of the human power to disrupt the very nature within which our lives are embedded. There is a special class of issues that science and religion must jointly address. They emerge at the leading edge of environmental issues where fact diminishes, uncertainty looms large, and value judgment is needed. Permit me to use the work of our research group on technological hazards to illustrate and let me apologize beforehand that my data are primarily from the United States.

COPING WITH HAZARDS: A CASE STUDY

In a recent U.S. poll, 78 per cent of the public surveyed agreed that 'people are subject to more risk today than they were twenty years ago,' and only 6 per cent agreed that there was less risk. Further, some 55 per cent agreed with the statement that 'the risks to society from serious scientific and technological advances will be somewhat greater 20 years from now than they are today,' while only 18 per cent felt the risks will be somewhat less (Harris 1980).

As a scientist with the factual data at my command, I can't support this view but the basis for these perceptions is easily understood. There is literally a 'hazard of the week' in the media and some forty to fifty new or newly discovered hazards are discussed annually in leading newspapers and periodicals (Kates 1977). Nuclear weaponry proliferates, there are some hundred thousand chemicals in commerce, as there are some twenty thousand separate consumer products. And added to these technological hazards are the social hazards of crime and unemployment.

The burden of technological hazard, the subject of our research group's analysis, is considerable. In the United States alone, the estimated social costs of hazards associated with the manufacture and use of technology, including property damage, losses of productivity from illness or death, and most but not all of the costs of control, amounted, in 1979, to between $179 and $283 billion (equivalent to 7.8 to 12.4 per cent of GNP; Tuller, forthcoming). Similarly, 15 to 25 per cent of annual human mortality is associated with various technologies. No similar analyses have been done elsewhere in the world, but there is every reason to believe that similar patterns of damage and loss would be found in most industrialized countries and in certain areas of developing countries where they might even be greater.

In contrast to social cost and human mortality, the ecosystem impacts, while conceivably the most important in the long term, are difficult to quantify. Death certificates are not filled out for the millions of non-human living organisms related to each other by complex chains of interdependence. At best, only a few crude indicators are available. One indicator is species extinction, particularly of birds and mammals. Current rates of known extinction of bird and mammalian species (slightly less than one per year) are estimated to be ten to fifteen times larger than those at the beginning of the industrial revolution. About one-half of these extinctions appear to be directly related to technology (Harriss, Hohenemser, and Kates 1978). If we include all of the estimated two to twenty million (average ten million; Ehrlich and Ehrlich 1981) existing species of all forms of life, then it is estimated that one million species will be extinct by the end of the century, at a rate of one a day currently and one an hour at the end of the period (Myers 1979).

A second measure of ecosystem impact is productivity or the ability of ecosystems to produce organic material from inorganic materials and sunlight. Using as an indicator of productivity the changing magnitude of the land biomass, recent estimates (made with great uncertainty) indicate a net annual decline of 0.2 to 2 per cent in global land mass, about three-quarters of which is technologically related (Woodwell et al. 1978). Other indicators are the world-wide declines in natural resource productivity, marked by rising real prices and declines in production of forest lumber and fish (Ridker and Watson 1980).

None the less, there is cause for cautious optimism regarding some technological hazards. There is surely a time lag between the employment of new technologies and the identification of their hazardous impacts, but the period seems to be decreasing (Kasperson 1977). For most acute effects and for some chronic ones as well, there is either improvement or at least no sign of worsening. Accidental death rates have declined in the United States despite increased use of technological devices, and significant improvement has been made in reducing the concentration of three of five major pollutants in the air. Thus for technologies whose consequences are immediate, whose sources are identifiable, and whose mechanisms are reasonably well understood, the problem of achieving a tolerable level of risk, though expensive and challenging, is acceptable, a concomitant to the widespread beneficial use of modern technology.

More difficult to deal with are acute hazards of a rare catastrophic nature, or slow and cumulative hazards of great persistence, spread, or toxicity. In a recent comparison of ninety-three technological hazards we have identified five sets of hazards and independent underlying dimensions, four of which represent types that are difficult to cope with. These four are: intentional biocides, the persistent, delayed teratogens, the rare catastrophes, and diffuse global threats. The fifth group, which we call the common killers, is composed of such hazards as automobiles or household drugs.

The *intentional biocides* derive their lethality from the great toxicity innate in the design of technologies meant to hurt living organisms: humans (in the case of weapons), insects (pesticides), vegetation (herbicides and chainsaws), bacteria and viruses (drugs). Highly efficient, these technologies are usually narrowly targeted and access to them is restricted. But if by error of design or application or by intention they drift off target, they are enormously dangerous. Thus in our time we have already experienced nuclear fallout, massive fungicide poisoning in Iraq, and world-wide antibiotic-resistant strains of venereal disease, and these threats are still increasing.

Persistent, delayed teratogens and mutagens comprise a class of hazards whose inherent danger arises from a combination of characteristics each of which though threatening is manageable by itself. It is the combination of long life for the material, long delay until consequences appear, and transgenerational impact that makes them so hazardous. Long-lasting metals such as antimony, cadmium, lead, mercury, molybdenum, and selenium have man-made fluxes between five and eighty times those found in nature; all of these elements persist almost forever and cause serious disruption to living organisms; they can accumulate slowly in the biosphere and several are mutagenic (Harriss and Hohenemser 1978). Heavy metals and radiation, the other major source of persistent mutagens, have already caused numerous poisoning episodes and a few major epidemics such as one of Minimata disease and the seed grain mass poisoning in Iraq.

For most of the *rare catastrophes* the mechanisms are well understood: a jumbo jet collision, the collapse of a dam, a nuclear reactor accident, a liquid natural gas (LNG) tank explosion, or the fall of a satellite. We have already experienced a number of major dam failures, one jumbo jet collision, a single LNG tank explosion, and near misses with satellites and nuclear reactors. For some catastrophes, such as the accidental creation of a virulent micro-organism through hybrid or recombinant DNA technology, the threat while real is almost impossible to assess reliably or perhaps even to verify if it did occur.

Finally there are the *diffuse global threats* caused by materials disbursed world-wide that slowly but steadily accumulate, mainly in or through the atmosphere, and threaten to change the climate, destroy the protective ozone layer, or increase the acidity of precipitation. The threats are identifiable but much scientific uncertainty still remains regarding the speed at which they are evolving, their sources, the mechanisms involved, and their impacts.

Note that to date each of these four groups of hazards is threatening in its potential rather than actual record of human and ecosystem injury. Each group poses a different sort of threat based on very different criteria of intrinsic hazardousness. And, finally, we have studied only ninety-three hazards involving several hundred individual technologies, but the list of hazardous components, products, or processes numbers in the hundreds of thousands.

Issues for science and religion

Do we need to do everything we do because it is to someone's advantage or do we do it simply because we can? Is all of this large burden of technological hazard necessary? Where the harm of a technology clearly outweighs its value to society, the answer is 'no.' We no longer manufacture thalidomide or legally hunt the blue whale. But it is seldom that the balance between harm and value can be simply and clearly struck. Thus we use DDT and SSTs in some countries and not in others, and ban the pesticide Leptophos in the United States and not in Canada.

We strike these different balances for a variety of reasons to do with different societal needs, risks, and influences, but also because of an inherent uncertainty about how harm occurs and how much technologies are worth. A new quasi-discipline of risk–benefit analysis has arisen to deal with such questions and thereby poses the first of the issues that exemplify matters of joint concern.

'*... and How Much for Your Grandmother*?' (Adams 1974)

Inherent in most societal choices about coping with hazards is the valuation of a human life. Despite the argument that life is priceless and can never be valued, both explicit and implicit valuations are being made all the time. Juries and courts give awards to heirs for the loss of life. My university currently values my life at two times my annual salary, and more if I die accidentally.

More common, however, is the implicit valuation that takes the form of expenditures to prevent life loss or to save an endangered life, or as compensation for increased risk (Graham and Vaupel 1981). The U.S. Nuclear Regulatory Commission is currently willing to require a utility to expend up to $5 million per life saved in measures employed to prevent radiation exposure, while the U.S. Department of Transportation is unwilling to order the expenditure of $13,000 per life saved through requiring the use of airbags.

Two major techniques for setting a price on a life are currently employed: willingness-to-pay and the human capital approach. In the first instance, life or at least life-saving is viewed as though it were a commodity in a market in which consumers or decision makers are willing to spend or receive various amounts for life-saving or life-risking activities. Life is worth what the market – be it economic or political – is willing to pay. In the second case, life is worth what it can produce, and lost earnings from premature death are the basic measure of valuation.

The issue of life valuation is only one of a family of related issues. Some are issues of equality – are lives in the future less valued than lives today? Some are issues of perception – are many people dying together valued more than the same number dying singly? or is a cancer death more dreadful than a drug overdose? All are issues that should encourage our collaboration. Human life is too valuable to be valued only by those who find it easy to do so.

'Would we miss the snail darter?' (Ehrlich and Ehrlich 1981)

If the valuation of human life poses knotty questions to both science and religion, then the valuation of non-human life may be even more troubling, bringing us into direct confrontation with our relationship to the natural world. At any moment of time it is difficult to be sanguine about the valuation of human life: the memory of the Holocaust is still too fresh, the threat of nuclear holocaust too real, the existence of capital punishment too common. Yet clearly we are at a moment in time when a pervasive effort is being made to extend our evaluations of human life to the life of other organisms: to the human foetus, to animals, and to other species.

Human population increase and technology have vastly accelerated the extinction of species. The gentle balance between new species evolution and old species extinction which seemed to favour overall the proliferation of species is being reversed (Myers 1979; Ehrlich and Ehrlich 1981). An extinction of species, the biological doomsayers' claim, as great as the disappearance of the dinosaurs sixty million years ago, is at hand.

There are four major arguments for the preservation of species (Ehrlich and Ehrlich 1981): ethical, esthetic, economic, and ecological. Other species should be preserved because they have a right to existence, because they are beautiful, because their sustained yield is valuable, and because they are necessary to human life-support systems. None of the utilitarian arguments (economic, ecological) seem sufficiently compelling to protect the entire range of threatened life, particularly insects, fish, and plants. Only a serious extension of human rights to non-human life can slow the rate of human-induced extinction.

The abominations of Prometheus

Balancing risks and benefits is one way of coping with hazards; averting them is another. The fine balancing exemplified by life valuation may be beyond our skill or morality and the measure of harm may be beyond our science. In the face of great uncertainty, most societies employ some measure of aversion simply to reduce the anxiety they must cope with. In short, they rely on taboos. All cultures contain taboos both sacred and profane and the most universal taboos relate to the most elementary needs: life, sex, food.

Taboos are easily recognized as such in so-called primitive societies, but are made sacred when they are part of the world's great religions and are ignored when embedded so deeply in one's native culture as to constitute habit or law. Thus 'primitive' efforts to avoid the risk of defilement are labelled as taboo, but similar efforts in our culture to avoid the risk of disease are called preventive medicine. Or in many countries where organized killing is sanctioned (war, capital punishment), self-killing is prohibited and the suicide taboo is part of the law.

There are a few recognized taboos for hazards: it is illegal in the United States to add a carcinogen to food (although legal to add one to air or water), or to eliminate an entire species of living things through the development of a large project such as a dam or highway. In both cases the decision to give the taboo the force of law was not made by weighing costs and benefits; rather it was a simple imperative – thou shall not add carcinogens to food or eliminate species with government projects.

Taken by themselves out of the context of the whole, taboos don't necessarily make sense. Our need for technological taboos rests not on our inability to cope with any specific hazard but on our inability to cope collectively with all of them. Drawing a line by preserving the snail darter endangered by the Tellico Dam in Tennessee may be marginally meaningless, but drawing a line somewhere may be fundamental to human survival. Science can help in selecting candidates for aversion by identifying particularly hazardous technologies or behaviour, but credible taboos require reference to higher principles.

Three-dimensional justice

The hazard burden is large in the aggregate but is manageable if shared fairly. Unfortunately, the distribution of risks and that of benefits from most technologies are not concordant; either risk or benefit may be concentrated and the other diffuse. The family of issues that arise from these discrepancies is called equity or fairness issues. In our own research we have been exploring the issue of equity or fairness in three contexts: locus, legacy, and labour–laity. The first inquires about geographical equity – in whose backyard will the noxious fumes or traffic be found? The second considers temporal equity and pursues the problem of intergenerational justice for long-lasting risks, primarily hazardous wastes and the risk of resource or technology scarcity. The third relates to social class, in this case workers and their occupational hazards as compared with the general public which is uninvolved in production or the use of a specific technology. We have studied the first two equity issues primarily in the context of the disposal of radioactive waste, and by extension of all long-lived hazardous waste.

There is widespread technical optimism that radioactive wastes can be effectively managed, that is, assembled, processed, and stored in ways that limit the risk to levels well below that of naturally occurring radiation (Aikin, Harrison, and Hare 1977). However, except for one proposal to bury high-level radioactive waste in equally spaced holes drilled in almost everybody's backyard (Cohen 1977; Ringwood 1980), all proposals to deal with such waste involve concentrating the risk on an area and population considerably smaller than would benefit from the nuclear power production. And from this principle emerges a wide range of social and political problems that have effectively outdistanced progress towards a technical solution (R. Kasperson 1979). Faced with the pressing need for storage of radioactive and other hazardous wastes, there is increasing recognition of these equity issues and some modest exploration of both compensatory and procedural mechanisms for achieving greater fairness.

For radioactive wastes, at least, there is also widespread popular recognition of the intertemporal equity problem; the long half-life of radioactive fission products and actinides is an active subject in the nuclear debate. Despite much lip service, however, future generations have little standing. In U.S. law at least there is hardly any precedent for providing such standing beyond the immediate generation (Green, forthcoming). While there have been novel suggestions for providing future generations with public advocates akin to guardians (Maynard et al. 1976), none have been implemented and few spokespersons for the future have been forthcoming. The U.S. Environmental Protection Agency has proposed a rule that specifically considers future generations, namely not to burden the future with risks exceeding those to the present generation, but it too has not become part of administrative law (U.S. Environmental Protection Agency 1978).

The equity issues between labour and laity is not even on the public agenda, even though we have found it a significant and widespread problem (Derr et al., 1981). In forty out of the ninety-three hazards we studied both workers and the general public are at risk. In 75 per cent of these cases, workers were exposed to hazards at concentrations ten times or more greater than were the public. This differential exposure is matched by differential protection. Occupational standards of protection are generally set at about the level of minimal observed harm, while standards for the public are ten to a thousand times below that level. Surprisingly, even in such countries as Sweden or the USSR where stricter occupational standards have been enacted, the differential between labour and laity still persists, albeit at a more restrictive level. At stake in such differential exposure is a substantial toll of life and livelihood. The toll from accidents is well known. In the United States, for example, 11.6 per cent of 103,000 accidental deaths in the period studied were work related, as were 7 to 15 per cent of 75,000,000 injuries. Much disputed is the toll of occupational illness and disease, which is surely substantial, including estimates of 1 to 38 per cent of 387,000 deaths due to cancer. Some 17.6 per cent of fifteen million disabled adults attribute their condition to work-related accidents or disease.

Since the differential is rarely noted, it is not often called into question. But four major principles of justice have been or could be invoked in its defence: (1) utility – whereby the sum of human betterment is greater because of the differential, (2) ability – whereby workers are more physically able, skilled, or trained to tolerate greater risk, (3) compensation – whereby workers are paid to bear risks, and (4) consent – whereby workers are informed of the risks and voluntarily accept them.

There are two aspects to these claims: are they true and are they just? As scientists we can examine their veracity – the economics of differential social cost, the relative vulnerability of workers and publics, the risk premium involved in compensation, or the information workers have about risk. But even if all the claims were true, the issue of justice remains. For example, in the United States we do not permit individuals to sell their kidneys (only to donate them). Why should we permit workers to sell fractions of their lives (or to have greater fractions taken from them)?

In all three types of equity issues the limits of science in clarifying the factual issue are quickly reached and issues of justice emerge. Ian Barbour (forthcoming) notes that unlike secular society, the Judaeo-Christian tradition gives standing to future generations or posterity. Equity issues of hazards and resources are prime claimants for a renewed science–religion effort.

CONCLUSION

In the chapter entitled 'The Sacramental Universe' in *Nature, Man and God*, William Temple finds in 'the principle in which belief in sacraments reposes ... a clue to the understanding of the relation of spirit to matter in the universe' (p. 486). To relate spirit to matter was a major goal of his, to somehow inject the 'life of the spirit ... characterized by determination by the good' into the physical world of mechanical forces and chemical compounds. In the absence of the spirit, he feared that 'the unity of man's life is broken; the material world, with all man's economic activity, becomes a happy hunting ground for uncurbed acquisitiveness, and religion becomes a refined occupation for the leisure of the mystical.'

In the consideration of nature, the dualism that Temple seeks to resolve is more complicated. The 'determination by the good' is complicated by the grant of dominion over nature and the material world of living systems is a biological world more complex than 'chemical compounds and mechanical forces.' Thus dominion, system, and mystery are intertwined and none is the unique domain of science or religion.

Now, no more than in the past, today's environmental issues should be informed by both the 'determination by the efficient' and the 'determination by the good.' In many fortunate cases the efficient may concur with the good and the choice is simplified. But in others, such as in the valuation of human life, the selective avoidance of hazardous technology, or issues of equity, the good may be less efficient. And in the issues of conflict between some humans and other life, the good has yet to be defined. If William Temple were alive today, wouldn't he be interested in these questions?

NOTES

1 This paper draws extensively on work prepared during a residential fellowship at the Woodrow Wilson International Center for Scholars, Washington, DC, and under research grants PRA79-11934, OSS77-16564, and OSS79-24516 of the U.S. National Science Foundation.

2 Reviewers have called my attention to several exceptions to this generalization, an example of which is the Religion/Environment project of the Sigurd Olson Environmental Institute at Northland College, Ashland, Wisconsin,

that 'seeks to explore the spiritual dimensions of caring for the earth, the concept of stewardship of nature, and kinship of all created beings.'

REFERENCES

Adams, J.G.U. 1974. '... and How Much for your Grandmother?' in S.E. Rhoads ed, *Valuing Life: Public Policy Dilemmas*, 135–46. Boulder, Colo.: Westview Press

Aikin, A.M.; Harrison, J.M.; and Hare, F.K. 1977. *The Management of Canada's Nuclear Wastes*. Ottawa: Energy, Mines and Resources Canada. Report ET 77-6

Barbour, Ian G. Forthcoming. 'Nuclear Energy and Future Generations,' in Warner Klugman ed, *Nuclear Power: Ethics and Public Policy*

– 1972. *Earth Might Be Free: Reflections on Ethics, Religion and Ecology*. Englewood Cliffs, NJ: Prentice-Hall

– ed. 1973. *Western Man and Environmental Ethics*. Reading, Mass.: Addison-Wesley Publishers

Bergonioux, F.M. 1961. 'Notes on the Mentality of Primitive Man,' in S.L. Washburn ed, *Social Life of Early Man*. New York: Wenner-Gren Foundation

Black, John. 1970. *The Dominion of Man*. Edinburgh: The University Press

Blanc, Alberto C. 1961. 'Some Evidence for the Ideologies of Early Man,' in S.L. Washburn ed, *Social Life of Early Man*. New York: Wenner-Gren Foundation

Blurton-Jones, Nicholas; and Konner, Melvin J. 1976. '!Kung Knowledge of Animal Behavior,' in R.B. Lee and I. De Vore eds, *Kalahari Hunters-Gatherers*. Cambridge, Mass.: Harvard University Press

Cohen, Bernard L. 1977. 'High-Level Radioactive Waste from Light-Water Reactors,' *Rev. Mod. Phys.* 49 (Jan.): 22

Conklin, Harold C. 1980. *Ethnographic Atlas of Ifugao*. New Haven: Yale University Press

Derham, William. 1798. *Physico-theology: Or, a Demonstration of the Being and Attribution of God, from His Works of Creation*. 2 vols. London: A. Strahan et al.

Derr, Patrick; Goble, Robert; Kasperson, Roger E.; and Kates, Robert W. 1981. 'Worker/Public Protection: The Double Standard,' *Environment* 23, no. 7 (Sept.), 6–15, 31–6

Douglas, Mary. 1966. *Purity and Danger: An Analysis of Concepts of Pollution and Taboo*. London: Routledge and Kegan Paul

– 1972. 'Deciphering a Meal,' *Daedalus, J. Acad. Arts and Sci.* (winter): 68–81

Ehrlich, Paul and Anne. 1981. *Extinction: The Causes and Consequences of the Disappearance of Species*. New York: Random House

Eiseley, Loren. 1958. *Darwin's Century*. Garden City, NY: Doubleday and Co.
Evans-Pritchard, E.E. 1965. *Theories of Primitive Religion*. Oxford: The Clarendon Press
Fraiberg, Selma H. 1968. *The Magic Years*. New York: Charles Scribner's Sons
Graber, Linda H. 1976. *Wilderness as Sacred Space*. Washington, DC: The Association of American Geographers
Glacken, Clarence J. 1967. *Traces on the Rhodian Shore*. Berkeley: University of California Press
Graham, John D.; and Vaupel, James W. 1981. 'Value of a Life: What Difference Does It Make,' *Risk Analysis* (March): 89–95
Green, Harold P. Forthcoming. 'Legal Aspects of Intergenerational Equity Issues,' in Roger E. Kasperson ed, *Equity Issues in Radioactive Waste Management*. Cambridge, Mass.: Oelgeschlager, Gunn and Hain
Harris, Louis and Associates. 1980. *Risk and a Complex Society*. Public opinion survey conducted for Marsh and McLennan. New York: Marsh and McLennan
Harris, Marvin. 1966. 'The Cultural Ecology of India's Sacred Cattle,' *Current Anthropology* 7: 51–9
– 1979. *Cultural Materialism: The Struggle for a Science of Culture*. New York: Random House
Harriss, Robert C.; and Hohenemser, Christoph. 1978. 'Mercury: Measuring and Managing the Risk,' *Environment* 20 (Nov.): 25–36
Harriss, Robert C.; Hohenemser, Christoph; and Kates, Robert W. 1978. 'Our Hazardous Environment,' *Environment* 20 (Sept.): 6–15, 38–44
Jackson, J.F. and J.H. 1978. *Infant Culture*. New York: Thomas G. Crowell
Johnson, Kirsten. 1977. 'Do as the Land Bids: Otomi Resource-use on the Eve of Irrigation.' Clark University unpublished PHD dissertation
Kaplan, Louise J. 1978. *Oneness and Separateness: From Infant to Individual*. New York: Simon and Schuster
Kasperson, Roger E. 1977. 'Societal Management of Technological Hazards,' in Robert W. Kates ed, *Managing Technological Hazard: Research Needs and Opportunities*, 49–80. Boulder: University of Colorado, Institute of Behavioral Science
– 1979. 'The Dark Side of the Radioactive Waste Problem,' in T. O'Riordan and R. d'Arge eds, *Progress in Resource Management and Environmental Planning*. New York: Wiley
Kates, Robert W. ed. 1977. *Managing Technological Hazard: Research Needs and Opportunities*. Boulder: University of Colorado, Institute of Behavioral Science
Leiss, William. 1972. *The Domination of Nature*. Boston: Beacon Press
Lovejoy, Arthur; and Boas, George. 1935. *Primitivism and Related Ideas in Antiquity*. Baltimore: The Johns Hopkins Press

McHarg, I.L. 1969. *Design with Nature*. New York: Natural History Press

Maimonides, Moses. 1881. *The Guide for the Perplexed*. London: Routledge

Maynard, William S. et al. 1976. *Public Values Associated with Nuclear Waste Disposal*. Seattle: Battelle Memorial Institute

Miller, James Grier. 1978. *Living Systems*. New York: McGraw-Hill

Myers, Norman. 1979. *The Sinking Ark: A New Look at the Problem of Disappearing Species*. Oxford: Pergamon

National Council of Churches of Christ. 1974. *Energy and Ethics: The Ethical Implication of Energy Production and Use*. New York: Energy Education Project, NCCC

– 1975. *The Plutonium Economy*. New York: NCCC

National Wildlife Federation (The). 1978. *Conservation Directory*. Washington, DC: National Wildlife Federation

Page, W.; and Richards, P. 1977. 'Agricultural Pest Control by Community Action: The Case of the Variegated Grasshopper in Southern Nigeria,' *African Environment*, 2/4 and 3/1 (Nov.): 127–41

Passmore, John. 1974. *Man's Responsibility for Nature*. New York: Charles Scribner's Sons

Ptai, Raphael. 1978. 'Comments on Ecology, Evolution and the Search for Cultural Origins: The Question of Islamic Pig Prohibition,' *Current Anthropology* 14: 518–21

Rappaport, Roy A. 1968. *Pigs for the Ancestors: Ritual in the Ecology of a New Guinea People*. New Haven: Yale University Press

Ray, John. 1759. *The Wisdom of God Manifested in the Works of Creation*. London: Rovington, Ward, Richardson

Ridker, Ronald G.; and Watson, William D. 1980. *To Choose a Future: Resource and Environmental Consequences of Alternative Growth Patterns*. Baltimore: Johns Hopkins Press

Ringwood, Ted. 1980. 'Safety in Depth for Nuclear Waste Disposal,' *New Scientist* 88, no. 1229 (27 Nov.): 574–5

Santmire, H. Paul. 1970. *Brother Earth*. New York: Thomas Nelson

Schaeffer, Francis A. 1970. *Pollution and the Death of Man*. Wheaton, Ill.: Tyndale House Publishers

Soler, Jean. 1979. 'The Dietary Prohibitions of the Hebrews,' *New York Review of Books*, 14 June 1979, 24–30

Tansley, A.G. 1935. 'The Use and Abuse of Vegetational Concepts and Terms,' *Ecology* 16: 287–307

Temple, William. 1953. *Nature, Man and God*. The Gifford Lectures, University of Glasgow, 1932–33 and 1933–34. London: Macmillan

Tuan, Yi-Fu. 1968. 'Discrepancies between Environmental Attitude and Behaviour: Examples from Europe and China,' *Canadian Geographer* 12: 176–91

Tuller, James. Forthcoming. 'Economic Costs and Losses,' in R.W. Kates and C. Hohenemser eds, *Technology as Hazard*. Cambridge, Mass.: Oelgeschlager, Gunn and Hain

United States. Environmental Protection Agency. 1978. 'Criteria for Radioactive Wastes: Recommendations for Federal Radiation Guidelines,' *Federal Register* 43 (15 Nov.): 53266

Von Bertalanffy, L. 1950. 'The Theory of Open Systems in Physics and Biology,' *Science* 111: 23–9

White, Gilbert. 1897. *Natural History and Antiquities of Selborne*. London: Macmillan

White, Lynn Jr. 1967. 'The Historical Roots of Our Ecological Crisis,' *Science* 155 (10 March): 1203–7

– 1973. 'Continuing the Conversation,' in I.G. Barbour ed, *Western Man and Environmental Ethics*. Reading, Mass.: Addison-Wesley

Williams, Raymond. 1976. *Keywords: A Vocabulary of Culture and Society*. New York: Oxford University Press

Woodwell, G.M. et al. 1978. 'The Biota and the World Carbon Budget,' *Science* 199 (Jan.)

Index

This book

was designed by

ANTJE LINGNER

of University of

Toronto

Press

BL
240.2
.E895
1983

359368